Student Solutions Manual for Rao's Statistical Research Methods in the Life Sciences

Charles D. Kincaid
Kellogg Company

Scott Morrison

Duxbury Press

An Imprint of Brooks/Cole Publishing Company

I(T)P® An International Thomson Publishing Company

Pacific Grove • Albany • Belmont • Bonn • Boston • Cincinnati • Detroit • Johannesburg • London
Madrid • Melbourne • Mexico City • New York • Paris • Singapore • Tokyo • Toronto • Washington

Sponsoring Editor: *Cynthia Mazow*
Marketing: *Michele Mootz*
Editorial Assistants: *Rita Jaramillo, Kimberly Raburn*
Production: *Dorothy Bell*

Cover Design: *Harry Voigt*
Cover Illustration: *Spike Walker/Tony Stone Images*
Printing and Binding: *Patterson Printing*

For more information, contact Duxbury Press at Brooks/Cole Publishing Company:

BROOKS/COLE PUBLISHING COMPANY
511 Forest Lodge Road
Pacific Grove, CA 93950
USA

International Thomson Editores
Seneca 53
Col. Polanco
11560 México, D. F., México

International Thomson Publishing Europe
Berkshire House 168-173
High Holborn
London WC1V 7AA
England

International Thomson Publishing GmbH
Königswinterer Strasse 418
53227 Bonn
Germany

Thomas Nelson Australia
102 Dodds Street
South Melbourne, 3205
Victoria, Australia

International Thomson Publishing Asia
60 Albert Street
#15-01 Albert Complex
Singapore 189969

Nelson Canada
1120 Birchmount Road
Scarborough, Ontario
Canada M1K 5G4

International Thomson Publishing Japan
Hirakawacho Kyowa Building, 3F
2-2-1 Hirakawacho
Chiyoda-ku, Tokyo 102
Japan

Printed in the United States of America
10 9 8 7 6 5 4 3 2 1
ISBN 0-534-93437-4

Contents

Chapter 1

Statistics: Its Objectives and Scope

1.1. There are two populations. The first is the population of tumor growth rates in rats treated with the drug. The second is the population of tumor growth rates in rats not treated with the drug.

1.3. There are four populations, one for each fertilizing method. Each population consists of all yields of all crops to which the fertilizing method is applied.

1.5. The two populations are the potential numbers of nematodes in soil around citrus trees three months after applications of each of the nematocides.

1.7. The population of interest is the potential collection of measurements each of which consists of a pair of values (age of a pregnant mother, birth-weight of her child) for all pregnant mothers.

1.9. For all the male and female fourth grade children place a 0 in the population if the child does not exhibit the symptoms of ADD and a 1 if the child does. The two populations consist of the potential collection of 0's and 1's for all male fourth grade children and all female fourth grade children.

1.11. The forester could select a sample of, for example, $n = 25$ 0's and 1's from the population given in Exercise **1.2**. This could be done by selecting 25 trees and recording a 0 if the tree is not diseased and a 1 if it is.

1.13. (a) To determine the study design we first specify the research question. The EPA wants to know how the average radiation levels compare inside and outside homes built on reclaimed lands for each type of home. There are four populations corresponding to the four types of homes. Each population is the collection of measurements each of which consists of a pair of values (inside and outside radiation levels) taken on all homes of the given type built on reclaimed land in Florida.

The second step is to choose the sample size. The methods for doing this will be discussed in Chapter 7.

The third step is to decide how the samples will be taken, i.e. the sampling plan. Again, this will be discussed later in Chapter 7.

(b) This design differs from the design in Part **a** in Exercise **1.13** by the populations under consideration. The research question is how do the average radiation levels compare inside and outside homes built on reclaimed lands? There is only one population, i.e., the collection of measurements taken on all homes built on reclaimed lands. Each measurement consists of the pair of values, radiation levels inside and outside the home.

The sample size and sampling plan will be discussed in later chapters.

Chapter 2

Describing Statistical Populations

2.1. Each of the two populations (treated and untreated) in Exercise **1.1** are

1. Univariate because each measurement consists of only one value.

2. Conceptual because the population pertains to growth rates of rats even into the future.

3. Infinite because it includes all possible growth rates in an interval of reasonable values above zero.

4. Quantitative because the growth rates are numerical values.

5. Continuous because the growth rates take all values in an interval above zero.

2.3. Each of the four populations of crop yields in Exercise **1.3** are

1. Univariate because the measurements in each population take on only one value.

2. Conceptual because the population consists of all crop yields given that method of fertilizing.

3. Infinite because the crop yields can take on an infinite number of values in some interval greater than or equal to zero.

4. Quantitative because the crop yields are on a numeric scale.

5. Continuous because the crop yields can take on any value greater than or equal to zero.

2.5. The two population described in Exercise 1.5 are each

1. Univariate because the measurement, i.e. the number of nematodes in soil around citrus trees three months after application of a nematocide, is a single value.

2. Conceptual because the entire population is non-existing at this time.

3. Infinite because the population potentially includes any positive integer.

4. Quantitative because the number of nematodes is numeric.

5. Discrete because the number of nematodes can be represented in a sequence: 0, 1, 2, 3, ...

2.7. The population in Exercise **1.7** is

1. Multivariate because each measurement is a pair of values; that is, the age of the pregnant mother and the birth weight of her child.

2. Conceptual because the population of measurements includes future values and, thus, does not exist right now.

3. Infinite because the values of birth-weight can take on any value in an interval greater than zero.

4. Quantitative because the pair of values can be measured on numeric scales.

5. Continuous because birth-weights can take on any value in a non-negative interval.

2.9. The population of 0's and 1's described in **1.9** is

1. Univariate because the measurements, 0 or 1, are single values.

2. Conceptual because the population includes potential future fourth grade males and females.

3. Infinite because the number of 0's and 1's is potentially infinite.

4. Qualitative because the 0's and 1's are meant as labels and not as numeric values.

5. Discrete because the the values can be enumerated $\{0, 1\}$.

2.11. (a) The probability that a randomly selected litter will have size less than four can be written in terms of Y as $Pr\{Y < 4\}$.

$$Pr\{Y < 4\} = Pr\{Y = 1\} + Pr\{Y = 2\} + Pr\{Y = 3\}$$
$$= f(1) + f(2) + f(3) = 0.2 + 0.3 + 0.3 = 0.8$$

(b) The probability that a randomly selected litter will have size three or more can be written as $Pr\{Y \geq 3\} = Pr\{Y = 3\} + Pr\{Y = 4\} = f(3) + f(4) = 0.3 + 0.2 = 0.5$.

2.13. (a)　　i. The probability that an observed value of Y is more than 18 can be written as

$$Pr\{Y > 18\} = Pr\{Y = 19\} + Pr\{Y = 20\} + Pr\{Y = 21\} + Pr\{Y = 22\}$$
$$= f(19) + f(20) + f(21) + f(22)$$
$$= 0.25 + 0.20 + 0.03 + 0.02 = 0.50$$

ii.

$$Pr\{17 \leq Y \leq 20\} = Pr\{Y = 17\} + Pr\{Y = 18\} + Pr\{Y = 19\} + Pr\{Y = 20\}$$
$$= f(17) + f(18) + f(19) + f(20)$$
$$= 0.20 + 0.25 + 0.25 + 0.20 = 0.90$$

iii.

$$Pr\{Y < 18\} = Pr\{Y = 15\} + Pr\{Y = 16\} + Pr\{Y = 17\}$$
$$= f(15) + f(16) + f(17) = 0.02 + 0.03 + 0.20 = 0.25$$

(b)　　i. $Pr\{Y > 18\} = 0.50$ can be interpreted as 50% of the population of numbers of leaves in a variety of tobacco plant are greater than 18. In other words, approximately 50% of the randomly selected plants will have more than 18 leaves.

ii. 90% of the population of numbers of leaves are between 17 and 20, inclusive, or 90% of the randomly selected plants will have between 17 and 20 leaves.

iii. 20% of the randomly selected plants will have less than 18 leaves.

2.15. The distribution in tabular form is

(y_1, y_2)	$f(y_1, y_2)$
(Young, Low)	0.01
(Young, Normal)	0.16
(Young, High)	0.03
(Middle Age, Low)	0.06
(Middle Age, Normal)	0.32
(Middle Age, High)	0.08
(Old, Low)	0.04
(Old, Normal)	0.16
(Old, High)	0.14

2.17. (a) Similar to Part **c** in Exercise **2.16** the relative frequency of Y_1 can be calculated by the following

$$f(\text{Young}) = f(\text{Young}, \text{Low}) + f(\text{Young}, \text{Normal}) + f(\text{Young}, \text{High})$$
$$= 0.01 + 0.16 + 0.03 = 0.20$$
$$f(\text{MiddleAge}) = f(\text{MiddleAge}, \text{Low}) + f(\text{MiddleAge}, \text{Normal}) + f(\text{MiddleAge}, \text{High})$$
$$= 0.06 + 0.32 + 0.08 = 0.46$$
$$f(\text{Old}) = f(\text{Old}, \text{Low}) + f(\text{Old}, \text{Normal}) + f(\text{Old}, \text{High})$$
$$= 0.04 + 0.16 + 0.14 = 0.34$$

(b) $f(\text{MiddleAge}) + f(\text{Old}) = 0.46 + 0.34 = 0.80$

2.19. (a) The probability that a randomly selected child will make the same number of errors on both tests is
$$f(0,0) + f(1,1) + f(2,2) = 0.04 + 0.24 + 0.14 = 0.42$$

(b) The probability that a randomly selected child will make more errors on Test 1 than on Test 2 is
$$f(1,0) + f(2,0) + f(2,1) = 0.09 + 0.04 + 0.06 = 0.19$$

(c) The probability that a randomly selected child will make no more than 1 error on each of the two tests is

$$f(0,0) + f(0,1) + f(1,0) + f(1,1) = 0.04 + 0.03 + 0.09 + 0.24 = 0.40$$

2.21. (a) Variety A is better from the point of view of increased yield. Its mean yield is 60 bushels per acre which is higher than the mean yield of Variety B which is 55 bushels per acre. Also, Variety A is likely to produce yields almost as large as 100 bushels per acre, but Variety B is not likely to be much bigger than 65 or so.

(b) Variety B produces yield that is less variable. Its frequency distribution is more concentrated around 55 bushels per acre.

2.23. (a) The mean is $\mu = 1(0.2) + 2(0.3) + 3(0.3) + 4(0.2) = 2.5$. On the average the litter size will be 2.5 animals.

(b) The population median is a number η such that no more than half the measurements have values less than η and no more than half the measurements have values greater than η. 2.5 is a median because $Pr\{Y < 2.5\} = 0.5$ and $Pr\{Y > 2.5\} = 0.5$.

2.25. (a) After deleting all measurements with $y = 60$ the new table is

y	10	11	12	13	14	15
$f(y)$	0.15	0.15	0.10	0.10	0.15	0.15

Because the sum of all distinct values in a frequency distribution must sum to one, the probabilities must be recalculated. The sum of the probabilities is now 0.80. The new probabilities are

$$f(10) = \frac{0.15}{0.80} = 0.1875 \qquad f(11) = \frac{0.15}{0.80} = 0.1875$$
$$f(12) = \frac{0.10}{0.80} = 0.1250 \qquad f(13) = \frac{0.10}{0.80} = 0.1250$$
$$f(14) = \frac{0.15}{0.80} = 0.1875 \qquad f(15) = \frac{0.15}{0.80} = 0.1875$$

Now $f(10) + f(11) + f(12) + f(13) + f(14) + f(15) = 1.0$.

(b) The mean of the population is

$$\mu = \sum yf(y) = 10(0.1875) + 11(0.1875) + 12(0.1250) + \ldots + 15(0.1875) = 12.5.$$

This is also the median because $Pr\{Y < 12.5\} = 0.5$ and $Pr\{Y > 12.5\} = 0.5$. Because both parameters are the same either one can serve as a reasonable measure of the location of the modified population.

(c) With $y = 60$ measurements in the population the mean is 22 and the median is 13. Without these the mean and median are both 12.5. The extreme value pulls the mean away from most of the measurements and thus it can be less desirable as a measure of the location of the unmodified caterpillar population. The median is not affected very much by the extreme values so it is still a reasonable measure of the center of the population.

2.27. For symmetric distributions the mean is the same as the median. However, as noted before, the mean is more sensitive to unbalanced extreme values than is the median. This unbalance occurs when the distribution is asymmetrical. Extremely large values, i.e. right skewed populations, pull the mean to the right of the median, and extremely small values, i.e. left skewed populations, pull the mean to the left of the median.

2.29. (a) The first quartile of a distribution is a value, $\eta_{0.25}$, such that no more than 25% of the measurements are less than $\eta_{0.25}$ and no more than 75% are more than $\eta_{0.25}$. From the distribution of the number of leaves per shoot 5% of the measurements are less than 17 and 75% are greater than 17.

But, also, 25% of the measurements are less than 18 and 50% are greater than 18. Further, for any number between 17 and 18, 25% of the measurements are less than that number and 75% are greater than that number. Therefore, any number between 17 and 18 inclusive is a first quartile of the distribution.

(b) Note that the distribution is symmetric. It is then easy to see that an argument very similar to the above will show that any number between 19 and 20, inclusive, is a third quartile.

2.31. (a) The first quartile is 26, because 25% of the measurements fall below and 75% of the measurements fall above.

The second quartile is 42 because $0.25 + 0.25 = 0.50$ or 50% of the measurements fall below 42 and $0.25 + 0.25 = 0.50$ or 50% of the measurement fall above 42.

The third quartile is 60 because 75% of the measurements fall below 60 and 25% fall above.

(b) Yes, it is possible. The distribution is skewed right so the mean is larger than the median.

2.33. (a) For this distribution the values that are two standard deviations from the population mean are $\mu - 2\sigma = 350 - 80 = 270$ and $\mu + 2\sigma = 350 + 80 = 430$. Therefore, by the Empirical Rule approximately 95% of the scores will be between 270 and 430.

(b) Three standard deviations from the mean is $\mu - 3\sigma = 350 - 120 = 230$ and $\mu + 3\sigma = 350 + 120 = 470$. According to the Empirical Rule almost all of the test scores will be between 230 and 470.

2.35. (a) From Exercise **2.32** $\mu = 120$ and $\sigma = 10$. Since $100 = 120 - 2(10)$ and $140 = 120 + 2(10)$, 100 and 140 are two standard deviations away from the population mean. Tchebycheff's rule says that at least 75% of the yields will be between 100 and 140 bushels per acre.

(b) $105 = 120 - 1.5(10)$ and $135 = 120 + 1.5(10)$ are 1.5 standard deviations from the population mean. By Tchebycheff's rule at least $1 - \frac{1}{(1.5)^2} = 0.56$ or at least 56% of the yields will be between 105 and 135 bushels per acre.

2.37. For this distribution

$$\mu = \sum yf(y) = 15(0.2) + 16(0.3) + \cdots + 22(0.2) = 18.5$$

$$\sigma^2 = \sum (y - \mu)^2 f(y)$$

$$= (15 - 18.5)^2(0.2) + (16 - 18.5)^2(0.3) + \cdots + (22 - 18.5)^2(0.2) = 1.69$$

$$\sigma = \sqrt{\sigma^2} = 1.3$$

$$\text{C.V.} = \frac{\sigma}{|\mu|} = \frac{1.3}{18.5} = 0.0703$$

2.39. (a) The distribution of Y_1 is

y_1	0	1	2
$f(y_1)$	0.17	0.55	0.28

Then,

$$\mu = \sum yf(y) = 0(0.17) + 1(0.55) + 2(0.28) = 1.11$$

$$\sigma^2 = \sum (y - \mu)^2 f(y)$$

$$= (0 - 1.11)^2(0.17) + (1 - 1.11)^2(0.55) + (2 - 1.11)^2(0.28) = 0.4379$$

$$\sigma = \sqrt{\sigma^2} = \sqrt{0.4379} = 0.6617$$

(b) The distribution of Y_2 is

y_2	0	1	2	3
$f(y_2)$	0.17	0.33	0.34	0.16

Then,

$$\mu = \sum yf(y) = 0(0.17) + 1(0.33) + 2(0.34) + 3(0.16) = 1.49$$

$$\sigma^2 = \sum (y - \mu)^2 f(y)$$

$$= (0 - 1.49)^2(0.17) + (1 - 1.49)^2(0.33) + (2 - 1.49)^2(0.34) + (3 - 1.49)^2(0.16)$$

$$= 0.9099$$

$$\sigma = \sqrt{\sigma^2} = \sqrt{0.9099} = 0.9539$$

(c) The covariance is

$$\sigma_{12} = \sum (y_1 - \mu_1)(y_2 - \mu_2)f(y_1, y_2)$$

$$= (0 - 1.11)(0 - 1.49)(0.04) + (0 - 1.11)(1 - 1.49)(0.03)$$

$$+ (0 - 1.11)(2 - 1.49)(0.06) + (0 - 1.11)(3 - 1.49)(0.04)$$

$$+ (1 - 1.11)(0 - 1.49)(0.09) + \cdots + (2 - 1.11)(3 - 1.49)(0.04) = 0.0261$$

2.41. (a) Defining Y as the number of the seven patients who respond to the new drug, then Y has two outcomes with "does respond to the drug" as the success and "does not respond to the drug" as the failure. If the drug trials are identical, if the patients react to the drug independently of one another, and if the probability of a patient responding to the drug is constant, 0.70, for all patients, then Y can be regarded as a binomial random variable.

(b) Under the conditions in Part **a** in Exercise **2.41**, the probability that at least **five** out of the seven patients will respond to the test is

$$Pr\{Y \geq 5\} = f(5) + f(6) + f(7)$$

$$= \frac{7!}{5!2!}(0.7)^5(0.3)^2 + \frac{7!}{6!1!}(0.7)^6(0.3)^1 + \frac{7!}{7!0!}(0.7)^7(0.3)^0$$

$$= 0.3176 + 0.2471 + 0.0824 = 0.6471$$

2.43. Let Y be the number of revertants that the compound will produce.

(a) For this Poisson random variable with $\lambda = 9$

$$f(0) = \frac{e^{-9}9^0}{0!} = 0.0001$$

(b)

$$f(3) = \frac{e^{-9}9^3}{3!} = \frac{(0.0001)(729)}{6} = 0.0150$$

(c) First, recall that the sum of probabilities of all distinct values must be one. Then, the probability that Y is greater than five is one minus the probability that Y is less than or equal to 5, i.e.

$$Pr\{Y > 5\} = 1 - Pr\{Y \leq 5\} = 1 - [f(0) + f(1) + f(2) + f(3) + f(4) + f(5)]$$

Then,

$$f(0) = \frac{e^{-9}9^0}{0!} = 0.0001 \quad f(1) = \frac{e^{-9}9^1}{1!} = 0.0011$$

$$f(2) = \frac{e^{-9}9^2}{2!} = 0.0050 \quad f(3) = \frac{e^{-9}9^3}{3!} = 0.0150$$

$$f(4) = \frac{e^{-9}9^4}{4!} = 0.0337 \quad f(5) = \frac{e^{-9}9^5}{5!} = 0.0607$$

and

$$Pr\{Y > 5\} = 1 - [0.0001 + 0.0011 + 0.0050 + 0.0150 + 0.0337 + 0.0607]$$

$$= 1 - 0.1156 = 0.8844$$

2.45. (a) Assumptions that justify treating Y as a Poisson random variable are

    i. The probability of finding a caterpillar in an infinitesimally small area of the tree shoots is proportional to the area.

    ii. The probability of finding more than one colony in an infinitesimally small area of the tree shoots is negligible.

    iii. The occurrences of caterpillars in the different areas of the tree shoots are independent events.

(b) $\lambda = 2$ means that, on the average, two caterpillars occupy the shoots of a tree.

(c) For a Poisson random variable

$$f(y) = \frac{e^{-\lambda}\lambda^y}{y!}$$

Hence,

$$Pr\{Y = 0\} = f(0) = \frac{e^{-2}2^0}{0!} = 0.1353.$$

Thus, we can expect about 14 out of 100 trees to have no caterpillars.

2.47. Let Y be the random variable denoting the logarithm of the survival times of patients treated for a particular type of cancer.

(a) The proportion of treated patients that will survive more than two years is the area under a $N(1.1, 0.04)$ distribution within the interval $(2, \infty)$. The z-score for the endpoint of this interval is

$$z = \frac{2 - 1.1}{0.2} = 4.5$$

From Table C.1 the area of interest is less than 0.001. Therefore, the proportion of treated patients that will survive more than two years is less than 0.001.

(b) The z-score corresponding to the endpoint of 1 is

$$z = \frac{1 - 1.1}{0.2} = -0.5$$

By the symmetry of the normal distribution the area to the left of -0.5 (or less than 1 in the original units) is equal to the area to the right of 0.5. From Table C.1, this area is 0.3085. Thus, the proportion of treated patients that will survive less than one year is 0.3085.

(c) For a $N(1.1, 0.4)$ distribution, we want a value, t_0, such that the area less than t_0 under the curve is 0.90. This means that the tail area is 0.10. Table C.1 can be used "inside-out" to find the z-score. We look for a probability that is closest to 0.10 in the inside of the table. The z-score that corresponds to this entry is the desired z-score. From Table C.1, 0.1003 is the value closest to 0.10. The corresponding z-score is 1.28. Now, t_0, is calculated as

$$t_0 = \mu + 1.28\sigma = 1.1 + 1.28(0.2) = 1.356$$

Thus, 90% of the treated patients will survive less than 1.356 years.

2.49.

$$Y \geq \mu + z\sigma$$
$$\text{if and only if } Y - \mu \geq \mu + z\sigma - \mu$$
$$\text{if and only if } Y - \mu \geq z\sigma$$
$$\text{if and only if } \frac{Y - \mu}{\sigma} \geq z$$

2.51. First, we create the following table

y	$f(y)$	Prop. of values that are at least y
0	0.0012	1.0000
1	0.0124	0.9988
2	0.1240	0.9864
3	0.3421	0.8624
4	0.2422	0.5203
5	0.1231	0.2781
6	0.1403	0.1550
7	0.0147	0.0147

(a) For this distribution, $\alpha = 0.155$ is the only α−level that corresponds to a probability in the third column. Thus, of these five α−levels only $\alpha = 0.155$ has a corresponding critical value.

(b) The 0.155-level critical value is 6.

2.53. Section 2.2 in the companion text by Younger gives the code to do this problem.

Chapter 3

Statistical Inference: Basic Concepts

3.1. There is no way one can select a truly random sample. All one can do is to insure that the experimental rats represent, as well as is practically feasible, the population of rats being studied.

3.3. The sampling could be conducted in stages for convenience. The size of the region would determine the number of stages. For example, if the region were a single subdivision, then the first stage could randomly select a block in the subdivision. The second stage would randomly select a house on the block.

3.5. One possible method is the following. Obtain a list of all social security recipients in the county from the Social Security office. Number the recipients from 1 to N. Use a computer to generate random numbers from 1 to N and include the corresponding recipients in the sample.

3.7. (a) Class intervals of width 500 are chosen to create the frequency distribution. If the intervals were wider, then there would be too few. If the intervals were narrower, then there would usually be only one data point in each interval. With these classes the two frequency distributions are

	Supplemental		Control	
Class Interval	f_i	p_i	f_i	p_i
$9500 < y \leq 10000$	0	0.0000	3	0.2143
$10000 < y \leq 10500$	1	0.0714	0	0.0000
$10500 < y \leq 11000$	0	0.0000	6	0.4286
$11000 < y \leq 11500$	5	0.3571	0	0.0000
$11500 < y \leq 12000$	0	0.0000	2	0.1428
$12000 < y \leq 12500$	2	0.1428	1	0.0714
$12500 < y \leq 13000$	4	0.2857	1	0.0714
$13500 < y \leq 14000$	1	0.0714	1	0.0714
$14500 < y \leq 15000$	1	0.0000	0	0.0000

The histograms for these distributions are in the following figure

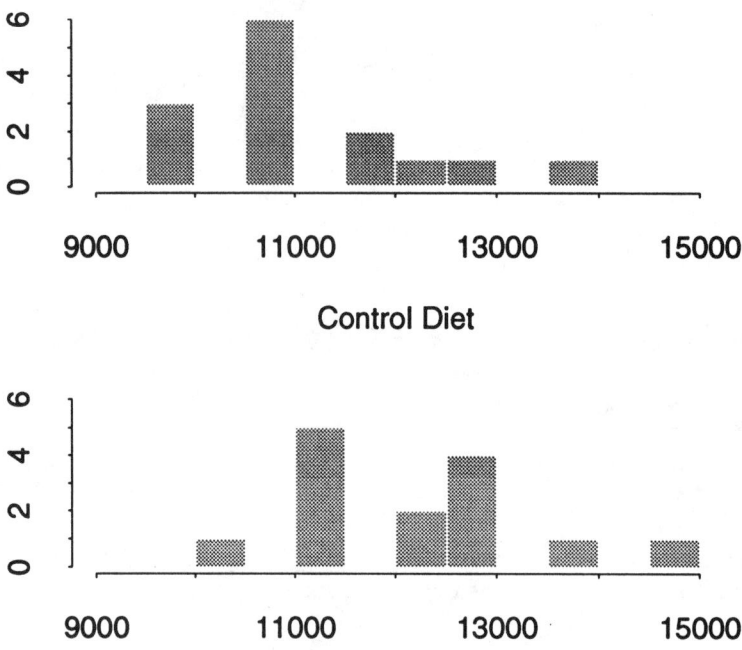

Control Diet

Supplemented Diet

(b) i. With such a small sample size it is difficult to really judge the normality of the population. However, the milk production of cows on both diets appear to be skewed a bit towards the right.

ii. The means do appear to be different. The distribution of total milk production for cows given the Supplemented Diet is shifted to the right of the distribution for cows given the Control Diet.

iii. The population variances do not appear to be too different. The distributions are spread out about equal amounts.

(c) The means and standard deviations for the two samples are

	Control	Supplemented
Mean	11180	12130
Standard Deviation	1214.461	1214.107

Comparing the mean milk production under the two diets, the mean milk production for the Supplemented diet is larger than the mean milk production for the Control diet. Also, the standard deviations for the milk production under the two diets are almost the same. Both of these conclusions are consistent with the conclusions based on visual comparison.

(d) i. The histograms were given above. The boxplots are

14

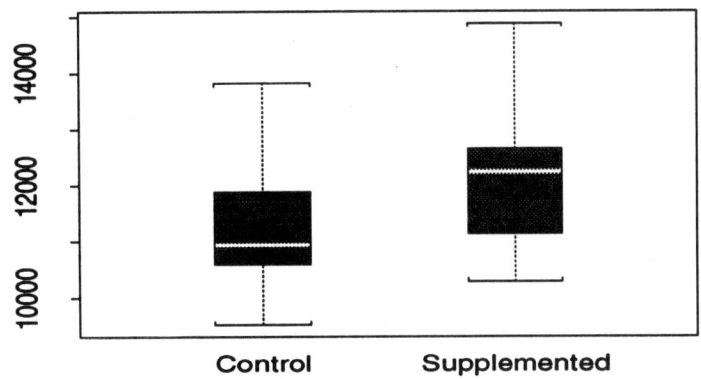

The stem-and-leaf plots of the data are, from S-plus,
Decimal point is 3 places to the right of the colon

```
 9 : 578
10 : 6899
11 : 0069
12 : 48
13 : 8
```
Decimal point is 3 places to the right of the colon

```
10 : 3
11 : 01134
12 : 225668
13 : 7
14 : 9
```

ii. The summary statistics from S-plus are

```
      Control        Supplemented
Min.    : 9518   Min.    :10280
1st Qu.:10630    1st Qu.:11160
Median :10930    Median :12240
Mean   :11180    Mean    :12130
3rd Qu.:11820    3rd Qu.:12640
Max.   :13820    Max.    :14870
Std Dev:1214.5   Std Dev:1214.1
```

(e) The shapes and variability are similar for the two distributions. The boxplots and statistics do show us that the median is shifted low for the control group and high for the supplemented group.

3.9. The figure allows the four distributions to be compared in many ways. A few of the points of note are the following. From the figure the box-plots show that the children in the third school district did better as a whole than all the children in the second and fourth school districts and better than at least seventy-five percent of the children in the first school district.

The middle fifty percent (the box) of the distributions of scores for districts two and four are very similar. The median scores are similar as well as the first and third quartiles. However, the whiskers for district two are uneven and show that the scores are skewed towards the upper values. On the other hand, the whiskers for district four are more symmetric showing that the scores are more evenly distributed in the tails.

3.11. (a) The samples that can be drawn from the target population $\{0, 1, 2, 6, 7, 9\}$ are

Sample number	Sample values	Sample mean	Sample number	Sample values	Sample mean
1	0, 1, 2	1	11	1, 2, 6	3.00
2	0, 1, 6	2.33	12	1, 2, 7	3.33
3	0, 1, 7	2.67	13	1, 2, 9	4.00
4	0, 1, 9	3.33	14	1, 6, 7	4.67
5	0, 2, 6	2.67	15	1, 6, 9	5.33
6	0, 2, 7	3.00	16	1, 7, 9	5.67
7	0, 2, 9	3.67	17	2, 6, 7	5.00
8	0, 6, 7	4.33	18	2, 6, 9	5.67
9	0, 6, 9	5.00	19	2, 7, 9	6.00
10	0, 7, 9	5.33	20	6, 7, 9	7.33

From this the sampling distribution of $\overline{Y}$ can be determined.

$\overline{y}$	1	2.33	2.67	3.00	3.33	3.67	4.0	4.33
$f(\overline{y})$	0.05	0.05	0.10	0.10	0.10	0.05	0.05	0.05

$\overline{y}$	4.67	5.00	5.33	5.67	6.00	7.33
$f(\overline{y})$	0.05	0.10	0.10	0.10	0.05	0.05

The samples that can be drawn from the target population $\{4, 8, 12, 20, 24, 32\}$ are

Sample number	Sample values	Sample mean	Sample number	Sample values	Sample mean
1	4, 8, 12	8	11	8, 12, 20	13.33
2	4, 8, 20	10.67	12	8, 12, 24	14.67
3	4, 8, 24	12	13	8, 12, 32	17.33
4	4, 8, 32	14.67	14	8, 20, 24	17.33
5	4, 12, 20	12	15	8, 20, 32	20.00
6	4, 12, 24	13.33	16	8, 24, 32	21.33
7	4, 12, 32	16	17	12, 20, 24	18.67
8	4, 20, 24	16	18	12, 20, 32	21.33
9	4, 20, 32	18.67	19	12, 24, 32	22.67
10	4, 24, 32	20	20	20, 24, 32	25.33

From this table, the sampling distribution of $\overline{Y}$ is

$\overline{y}$	8	10.67	12	13.33	14.67	16	17.33
$f(\overline{y})$	0.05	0.05	0.10	0.10	0.10	0.10	0.10

$\overline{y}$	18.67	20	21.33	22.67	25.33
$f(\overline{y})$	0.10	0.10	0.10	0.05	0.05

(b) The three sampling distributions are displayed below.

16

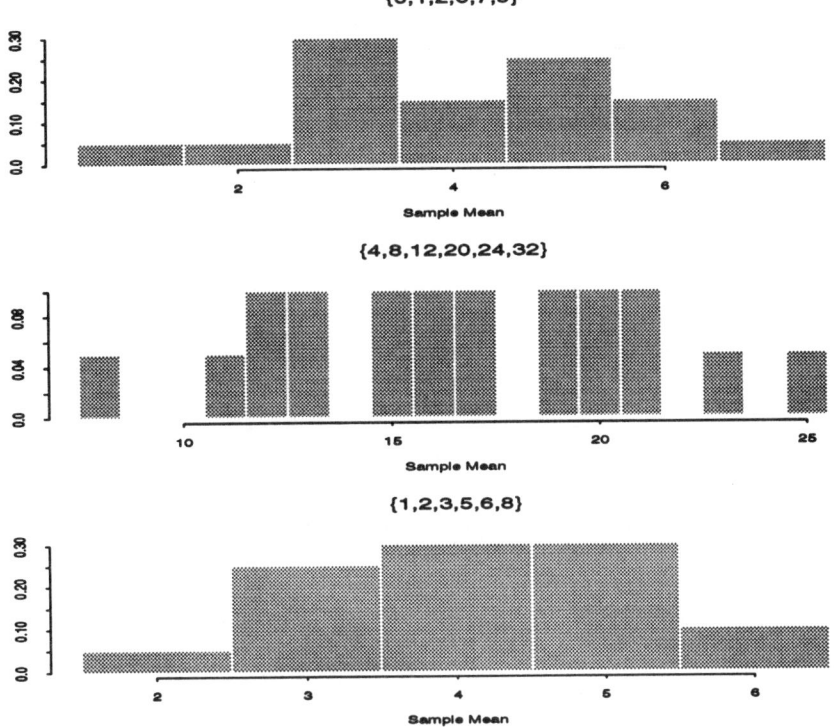

The first target population was not symmetric around its mean. The sampling distribution is also slightly skewed. The second target population is the most spread out and, likewise, the second sampling distribution is spread out and flat. Finally, the third target population is the most compact and symmetric. It's sampling distribution is the most compact and mound shaped.

(c) The mean and standard deviation of the first target population, $\{1,2,3,5,6,8\}$ were calculated as

$$\mu_{\overline{Y}} = 4.1667 \quad \text{and} \quad \sigma_{\overline{Y}} = 1.077.$$

For the second target population, $\{0,1,2,6,7,9\}$, the mean and standard deviation are

$$\mu_{\overline{Y}} = 4.1667 \quad \text{and} \quad \sigma_{\overline{Y}} = 1.492.$$

For the third target population, $\{4,8,12,20,24,32\}$, we have

$$\mu_{\overline{Y}} = 16.6665 \quad \text{and} \quad \sigma_{\overline{Y}} = 4.3097.$$

The population parameters for the original populations along with the parameters for the sampling distributions are

Population	μ	$\mu_{\overline{Y}}$	σ	$\sigma_{\overline{Y}}$
$\{0, 1, 2, 6, 7, 9\}$	4.1667	4.1667	2.4095	1.077
$\{1, 2, 3, 5, 6, 8\}$	4.1667	4.1667	3.3375	1.492
$\{4, 8, 12, 20, 24, 32\}$	16.6667	16.6667	9.6379	4.310

The population mean is the same for the original population and for the sampling distribution of $\overline{Y}$. The standard deviation for the original population is always larger than the standard deviation for the sampling distribution.

3.13. (a) The samples that can be drawn from the target population and the medians associated with the samples are

Sample number	Sample values	Sample median	Sample number	Sample values	Sample median
1	1, 2, 3	2	11	1, 5, 6	3.00
2	1, 2, 5	2	12	1, 5, 8	3.33
3	1, 2, 6	2	13	2, 5, 6	4.00
4	1, 2, 8	2	14	2, 5, 8	4.67
5	1, 3, 5	3	15	3, 5, 6	5.33
6	1, 3, 6	3	16	3, 5, 8	5.67
7	1, 3, 8	3	17	1, 6, 8	5.00
8	2, 3, 5	3	18	2, 6, 8	5.67
9	2, 3, 6	3	19	3, 6, 8	6.00
10	2, 3, 8	3	20	5, 6, 8	7.33

From this the sampling distribution of the sample median can be determined.

$\tilde{y}$ =median	2	3	5	6
$f(\tilde{y})$	0.20	0.30	0.30	0.20

(b) The mean and standard deviation of the sampling distribution of the sample median are, respectively

$$\mu_{\tilde{y}} = 2(0.20) + 3(0.30) + 5(0.30) + 6(0.20) = 4$$

and

$$\sigma_{\tilde{y}} = \sqrt{(2-4)^2(0.20) + (3-4)^2(0.30) + (5-4)^2(0.30) + (6-4)^2(0.20)} = 1.4832$$

(c) The mean of the target population is 4.1667. One unit away from this is 3.1667 and 5.1667. According to the sampling distribution of the sample mean, the probability that the sample mean is in this region is

$$f(3.33) + f(3.67) + f(4.0) + f(4.33) + f(4.67) + f(5.0) = 0.60.$$

From the sampling distribution of the sample median, the probability that the sample median is in this region is $f(5) = 0.30$.

The probability that the margin of error for the sample mean is within two units, or, in other words, that the sample mean is within $(2.1667, 6.1667)$ is

$$f(2.67) + f(3) + f(3.33) + \cdots + f(5.67) = 0.90.$$

The probability that the margin of error for the sample median is less than two is

$$f(3) + f(5) + f(6) = 0.80.$$

From these considerations the sample mean is a better estimator of the population mean than the sample median.

3.15. (a) Because the rainbow trout weights follow a normal distribution, the sampling distribution of the mean of the fifteen weights is normally distributed. The mean and standard deviation are

$$\mu_{\overline{Y}} = 1.5 \text{ pounds} \quad \text{and} \quad \sigma_{\overline{Y}} = \frac{0.4}{\sqrt{10}} = 0.1265 \text{ pounds}$$

(b) To find the probability that the mean of a random sample of $n = 15$ is within the interval $(\mu - 0.25, \mu + 0.25)$ first the z-scores of the endpoints are calculated

$$z = \frac{(\mu - 0.25) - \mu}{\sigma/\sqrt{n}} = \frac{-0.25}{0.1033} = -2.42$$
$$z = \frac{(\mu + 0.25) - \mu}{\sigma/\sqrt{n}} = \frac{+0.25}{0.1033} = +2.42$$

As in Chapter 2 we use Table C.1 to find the probabilities associated with these z-scores. Because of the symmetry of the normal distribution, the area in the tails greater than 2.42 is the same as the area less than -2.42. From the table, the tail area above 2.42 is 0.0078. Therefore, the entire area under the curve outside $(-2.42, 2.42)$ is $2(0.0078) = 0.0156$. Thus, the area under the curve within the interval is $1 - 0.0156 = 0.9844$. This means that the probability that the mean weight of a random sample of $n = 15$ rainbow trout will be within 0.25 pounds of the population mean weight is 0.9844.

3.17. (a) From Table C.2, $t(14, 0.001) = 3.7874$. Therefore, the probability that the observed value of t is larger than 3.79 (with roundoff) is 0.001.

(b)

$$t_c = \frac{\sqrt{n}(\overline{Y} - \mu)}{s} = \frac{\sqrt{15}(30 - 25)}{4} = 4.8412$$

(c) If the actual value of μ is 25, then we would expect to see a value greater than 3.79 about 0.1% of the time, which is very unlikely. Since we did observe a value this large either we observed a very rare event or the true value of μ is larger than 25. Because the event is so rare it is more likely that the true value is indeed larger than 25.

3.19. (a) The 0.001 level critical value of a $\chi^2(19)$−distribution is the $\chi^2(19, 0.001)$ value from Table C.3. In Table C.3 this is 43.8202.

(b)

$$\chi^2 = \frac{(20 - 1)10}{4} = 47.5$$

(c) Similar to Exercise **3.17**. The 0.001 level critical value is the value such that, if $\sigma^2 = 4$, then a randomly observed sample standard deviation will be larger than 43.8202 only 0.1% of the time. Because we did observe a value greater than the critical value 43.8202 and because the probability of this happening is so small if, in fact, $\sigma^2 = 4$, it is reasonable to conclude that, instead, $\sigma^2 > 4$.

3.21. (a) The 0.034-level critical value of a t-distribution with 13 degrees of freedom is 1.9904.

(b) The 0.84- and 0.06-level critical values of a χ^2-distribution with nine degrees of freedom are 4.9343 and 16.3459, respectively.

(c) The 0.75- and 0.12-level critical values of an F-distribution with six degrees of freedom for the numerator and eleven degrees of freedom for the denominator are 0.5654 and 2.2148, respectively.

3.23. (a) Estimation: The scientist wants to determine a set of plausible values for the proportion of children in this age group with ADHD. The population is the set of all zeroes and one's with a zero corresponding to a child without ADHD and a one corresponding to a child with ADHD. The parameter is the population proportion.

(b) Hypothesis Test: The scientist wants to verify the claim that the proportion of children in this age group with ADHD differs between males and females. The two populations are, first, all the zeroes and ones that correspond to all male children in this age group and, second, the zeroes and ones that correspond to all female children in this age group. The population parameters are the proportions in each population.

(c) Prediction: The scientist wants to predict the actual number of ones in a future sample of size 25. The population is the zeroes and ones of all future children of ages four through seven. Prediction problems do not focus on any specific parameter, rather on the random variable itself.

3.25. (a) In Exercise Part **c** in Exercise **3.11** the means of the sampling distributions were compared with the means of the original distributions. The following table summarizes the calculations.

Target	$\mu_{\overline{Y}}$	μ
1	4.1667	4.1667
2	4.1667	4.1667
3	16.6667	16.6667

Because the mean of the sampling distribution is equal to the mean of the original distribution this verifies that the estimator $\overline{Y}$ is unbiased.

(b) The MSE of $\overline{Y}$ is the mean of the population of all values of the squared differences, $(\overline{Y} - \mu)^2$. For the first target population, the first squared difference is calculated as $2.00 - 4.1667 = -2.1667$. After doing this for all sample means for each target population, the population of the squared differences for the three target populations can be tabulated as below. The mean of each sampling distribution is also calculated.

Target 1				
$u_1 = (\overline{y} - \mu)^2$	4.6946	2.2401	1.3612	0.7001	0.2467
$f(u_1)$	0.05	0.05	0.10	0.10	0.10
$u_1 = (\overline{y} - \mu)^2$	0.0278	0.0267	0.2533	0.6944	1.3533
$f(u_1)$	0.10	0.10	0.10	0.10	0.10
$u_1 = (\overline{y} - \mu)^2$	2.2599	4.6799			
$f(u_1)$	0.05	0.05			

The mean is

$$(0.05)4.6946 + (0.05)2.2401 + (0.10)1.3612 + \cdots + (0.05)4.6799 = 4.1665.$$

Target 2				
$u_2 = (\overline{y} - \mu)^2$	10.0280	3.3735	2.2401	1.3612	0.7001
$f(u_2)$	0.05	0.05	0.10	0.10	0.10
$u_2 = (\overline{y} - \mu)^2$	0.2467	0.0278	0.0267	0.2533	0.6944
$f(u_2)$	0.05	0.05	0.05	0.05	0.10
$u_2 = (\overline{y} - \mu)^2$	1.3533	2.2599	3.3610	10.0065	
$f(u_2)$	0.10	0.10	0.05	0.05	

The mean is

$$(0.05)10.0280 + (0.05)3.3735 + (0.10)2.2401 + \cdots + (0.05)10.0065 = 4.1665$$

Target 3				
$u_3 = (\overline{y} - \mu)^2$	75.1117	35.9604	21.7781	11.1336	3.9868
$f(u_3)$	0.05	0.05	0.10	0.10	0.10
$u_3 = (\overline{y} - \mu)^2$	0.4400	4.0132	11.1109	21.7464	36.0396
$f(u_3)$	0.10	0.10	0.10	0.10	0.10
$u_3 = (\overline{y} - \mu)^2$	36.0396	75.0528			
$f(u_3)$	0.05	0.05			

The mean is

$$(0.05)75.1117 + (0.05)35.9604 + (0.10)21.7781 + \cdots + (0.05)75.0528 = 16.6665.$$

Note that in each case the mean of the population of all values of the squared differences $(\overline{Y} - \mu)^2$, i.e. MSE, is the same as the variance of the sample mean, up to roundoff error. This equality holds because the sample mean is an unbiased estimator of μ.

3.27. The MSE as defined in the text is the mean of the population of all values of the squared difference, $(\hat{\theta} - \theta)^2$, between $\hat{\theta}$ and θ. That is, MSE $= \sum f(\hat{\theta})(\hat{\theta} - \theta)^2$.

If $\hat{\theta}$ is unbiased, then θ is the mean of the sampling distribution of $\hat{\theta}$. Therefore, the squared difference between $\hat{\theta}$ and θ are the squared deviations of the random variable $\hat{\theta}$ around its mean. The mean of squared deviations of a random variable around its mean is the variance of the random variable. Thus, for an unbiased estimator the MSE is the same as the variance.

3.29. (a) First,

$$\sigma_{\overline{Y}} = \frac{\sigma}{\sqrt{n}} = \frac{18}{\sqrt{15}} = 4.6476.$$

Then a 95% lower bound for μ is

$$\begin{aligned} \hat{\mu}_L &= \overline{Y} - z(\alpha)\sigma_{\overline{Y}} \\ &= 143 - 1.645(4.6476) \\ &= 135.3547 \end{aligned}$$

(b) A range of plausible values for the mean height of the population of all Douglas fir is $(135.3547, \infty)$. We have 95% confidence that the mean height of the population of all Douglas fir is greater than 135.3547.

(c) Yes, it is reasonable to say that the average tree height of the population is at least 130cm. All of the plausible values in the range calculated above, $(135.3547, \infty)$, are at least 130cm.

3.31. (a) It means that, with 99% confidence, the true fraction of the exposed population who suffer from the toxic effects of the chemical is at most 15%.

(b) No, it is not reasonable to conclude that more than 10% of the exposed population suffers from toxic side effects. Some of the values in the 99% confidence interval calculated above are more than 10%, but some of them are not.

It is reasonable to conclude that less than 20% of the exposed population suffers from toxic side effects. The entire range of plausible values is less than 20%.

3.33. (a) When testing H_{01} the research hypothesis states that the mean is more than 190. Thus, the research hypothesis is a statement about a lower bound for the population mean and H_{01} can be verified on the basis of a lower bound calculated from the data. On the other hand, when testing H_{02}, the research hypothesis concerns an upper bound for the population mean. We cannot verify this on the basis of a lower confidence bound.

(b) The calculated lower bound, 135.3547, is less than 190 cm. Therefore, there is evidence for the population mean height being no larger than 190 cm, as well as evidence for the population mean height being larger than 190 cm. Because there is evidence for both conclusions we cannot reject the null hypothesis.

(c) Using the confidence interval approach, the upper bound is first calculated. That is, with $\sigma_{\overline{Y}} = 4.6476$,

$$\hat{\mu}_U = \overline{Y} + z(\alpha)\sigma_{\overline{Y}} = 143 + 2.325(4.6476) = 153.8057$$

Because 153.8057 is larger than 100 there are values in the 99% confidence interval, $(-\infty, 153.8057)$, that support the null hypothesis H_{02} and values that don't support it. If the upper bound had been less than 100, then none of the evidence would have supported H_{02} and it would have been rejected. As it is, H_{02} can not be rejected.

3.35. (a) Consider $\mu_0 \leq \hat{\mu}_L = \overline{Y} - z(\alpha)\sigma_{\overline{Y}}$ so that we have

$$\mu_0 \quad \leq \overline{Y} - z(\alpha)\sigma_{\overline{Y}}$$
$$\text{if and only if} \quad z(\alpha)\sigma_{\overline{Y}} \quad \leq \overline{Y} - \mu_0$$
$$\text{if and only if} \quad z(\alpha) \quad \leq \frac{\overline{Y} - \mu_0}{\sigma_{\overline{Y}}}.$$

Therefore, if $\mu_0 \leq \hat{\mu}_L$ and the confidence interval approach rejects H_0, then $z = \dfrac{\overline{Y} - \mu_0}{\sigma_{\overline{Y}}}$ $\geq z(\alpha)$ will be true and the test statistic approach will also reject H_0 and vice versa. On the other hand, if $\mu_0 > \hat{\mu}_L$ so that the confidence interval approach does not reject H_0, then we also have $z = \dfrac{\overline{Y} - \mu_0}{\sigma_{\overline{Y}}} < z(\alpha)$ and the test statistic approach will not reject H_0 either.

(b) First, note that

$$\mu_0 \geq \quad \hat{\mu}_U = \overline{Y} + z(\alpha)\sigma_{\overline{Y}}$$
$$\mu_0 \geq \quad \overline{Y} + z(\alpha)\sigma_{\overline{Y}}$$
$$\text{if and only if} \quad -z(\alpha)\sigma_{\overline{Y}} \geq \quad \overline{Y} - \mu_0$$
$$\text{if and only if} \quad -z(\alpha) \geq \quad \frac{\overline{Y} - \mu_0}{\sigma_{\overline{Y}}}.$$

Then the argument follows the one above.

3.37. (a) At the 0.05 level, we would accept the alternative hypothesis of H_1: $\mu < 13$ if $\frac{\overline{Y} - \mu_0}{2/\sqrt{15}} < -1.645$. In other words

$$\frac{\overline{Y} - \mu_0}{\sqrt{2/15}} < -1.645 \Rightarrow \overline{Y} < 12.1505.$$

Therefore, for any given μ, the power of the test equals the area to the left of $13 - 1.645(2/\sqrt{15}) = 12.1505$ under a $N(\mu, 0.5164)$ distribution. This is equivalent to the area to the left of $\frac{12.1505 - \mu}{0.5164}$ in a $N(0,1)$ distribution. A statistics package such as SAS can be used to calculate the power for μ at the given levels. The power at these values of μ is

μ	12	11	10	9
Power	0.6147	0.9871	1.0000	1.0000

(b) The numbers calculated in Part **a** in Exercise **3.37** represent the probability of a 0.05 level test rejecting the null hypothesis for the given value of μ. For example, there is a 61% chance of the 0.05 level test saying that the population mean N-loss is less than 13% when the true population mean N-loss is actually 12%.

(c) See the plot in Part **d** in Exercise **3.37**.

(d) Replacing 1.645 by 2.326 in Part **a** in Exercise **3.37** gives the following table

μ	12	11	10	9
Power	0.3484	0.9391	0.9997	1.0000

The plot of the power at the 0.01 level and the 0.05 level is below.

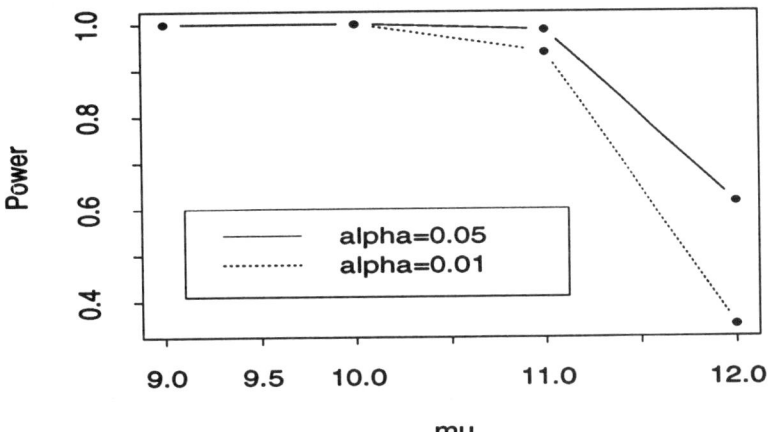

As can be seen in the graph, the power of the test at the 0.05 level is higher than at the 0.01 level. Performing the test at the 0.05 level rather than the 0.01 level means that we are willing to be more likely of making an incorrect rejection. We are more likely of making an incorrect rejection, because, in the long run, we are rejecting more often. Since we are rejecting more often we are also more likely to correctly reject.

3.39. First, note that, because the sampling distribution of $Y - \overline{Y}$ is a normal distribution with mean 0 and variance $\sigma^2 + \sigma^2/n$, $100(1-\alpha)\%$ of the observed values of $Y - \overline{Y}$ will be within the interval

$$\left(0 - z(\alpha/2)\sqrt{\sigma^2 + \frac{\sigma^2}{n}}, 0 + z(\alpha/2)\sqrt{\sigma^2 + \frac{\sigma^2}{n}}\right)$$

In other words,

$$Pr\left\{0 - z(\alpha/2)\sqrt{\sigma^2 + \frac{\sigma^2}{n}} \leq Y - \overline{Y} \leq 0 + z(\alpha/2)\sqrt{\sigma^2 + \frac{\sigma^2}{n}}\right\} = 1 - \alpha$$

or, again,

$$Pr\left\{\overline{Y} - z(\alpha/2)\sqrt{\sigma^2 + \frac{\sigma^2}{n}} \leq Y \leq \overline{Y} + z(\alpha/2)\sqrt{\sigma^2 + \frac{\sigma^2}{n}}\right\} = 1 - \alpha.$$

This last equality means that $100(1 - \alpha)\%$ of the intervals

$$\left(\overline{Y} - z(\alpha/2) \sqrt{\sigma^2 + \frac{\sigma^2}{n}}, \overline{Y} + z(\alpha/2) \sqrt{\sigma^2 + \frac{\sigma^2}{n}} \right)$$

will contain the future value of Y.

Therefore, a $100(1 - \alpha)\%$ prediction interval has the endpoints

$$Y_L = \overline{Y} - z(\alpha/2) \sqrt{\sigma^2 + \frac{\sigma^2}{n}}$$

$$Y_U = \overline{Y} + z(\alpha/2) \sqrt{\sigma^2 + \frac{\sigma^2}{n}}.$$

For the lower prediction bound, again note that, because the sampling distribution of $Y - \overline{Y}$ is a normal distribution with mean 0 and variance $\sigma^2 + \sigma^2/n$, $100(1 - \alpha)\%$ of the observed values of $Y - \overline{Y}$ will be greater than $0 - z(\alpha)\sqrt{\sigma^2 + \frac{\sigma^2}{n}}$. In other words, $100(1 - \alpha)\%$ of the time

$$Y - \overline{Y} \geq 0 - z(\alpha/2) \sqrt{\sigma^2 + \frac{\sigma^2}{n}}$$

or

$$Y \geq \overline{Y} - z(\alpha/2) \sqrt{\sigma^2 + \frac{\sigma^2}{n}}$$

Therefore, $Y_L = \overline{Y} - z(\alpha)\sqrt{\sigma^2 + \frac{\sigma^2}{n}}$ is a $100(1 - \alpha)\%$ lower prediction bound for Y.

3.41. (a) A 99% prediction interval for μ is

$$\left(\overline{Y} - z(0.01/2) \sqrt{(1 + 1/n)\sigma^2}, \overline{Y} + z(0.01/2) \sqrt{(1 + 1/n)\sigma^2} \right).$$

The endpoints are

$$\overline{Y} - z(0.005) \sqrt{(1 + 1/n)\sigma^2} = 2.84 - 2.575 \sqrt{(1 + 1/100)(0.7)^2} = 1.0285$$

and

$$\overline{Y} + z(0.005) \sqrt{(1 + 1/n)\sigma^2} = 2.84 + 2.575 \sqrt{(1 + 1/100)(0.7)^2} = 4.6515.$$

With 99% confidence, we predict that the GPA of a high school student selected at random from the population of high school students in the region will be between 1.0285 and 4.6515.

(b) The 99% confidence interval is $(2.822, 2.858)$ and the 99% prediction interval is $(1.0285, 4.6515)$. First, the prediction interval is wider because the variance of $Y - \overline{Y}$ is $\sigma^2 + \sigma^2/n$ which is larger than the variance of $\overline{Y}$.

Second, the purpose of the confidence interval is to enclose the population mean which is a fixed value. On the other hand, the purpose of the prediction interval is to enclose a random variable. There is much more uncertainty involved in predicting a random variable than in estimating a fixed value. The prediction interval must be wider because of this extra uncertainty.

3.43. (a) From the example, we have $\hat{\theta} = 2.57$ and $\sigma_{\hat{\theta}} = 0.75$. Next, the 90% confidence interval is

$$\left(\hat{\theta} - z(0.10/2)\sigma_{\hat{\theta}}, \hat{\theta} + z(0.10/2)\sigma_{\hat{\theta}} \right).$$

Therefore, the endpoints are

$$\hat{\theta} - z(0.05)\sigma_{\hat{\theta}} = 2.47 - 1.645(0.76) = 1.3198$$

and

$$\hat{\theta} + z(0.05)\sigma_{\hat{\theta}} = 2.47 + 1.645(0.76) = 3.8202$$

or

$$(1.3198, 3.8202).$$

(b) The null hypothesis is $H_0 : \theta = 0$ and the level of the test is 0.10 (because the interval is at 90% confidence). Because the interval does not contain 0, the data does support the conclusion that the true mean N-losses are not the same for the two fertilizers.

(c) To determine the $p-$value, we first calculate the test statistic.

$$z_c = \frac{2.57 - 0}{0.75} = 3.3816.$$

This is larger than any z-score in Table C.1. Thus, the $p-$value is less than $2(0.001)$ or $p-$value < 0.002. The smallest $\alpha-$level at which the test could be rejected is less than 0.002.

Chapter 4

Inferences About One or Two Populations: Interval Data

4.1. (a) Interval: The percentage change can be regarded as a point on a number line. The difference between two percentage changes provides a numerical measure of the amount by which the values differ. I.e., a percentage change of 30 and a percentage change of 20 differ by $30 - 20 = 10$.

(b) Ordinal: The values can be arranged by order of magnitude. *Low* is less than *Medium* which is less than *High*. However, the difference between two values gives no measure of the amount by which the statuses of two subjects differ.

(c) Nominal: The values of Type of Irrigation Method distinguish between distinct categories, but the categories cannot be ordered. It makes no sense to say that automatic sprinkler system is more than manual sprinkler system. It might make sense to say this if cost or convenience or efficacy were being measured, but then the variable would not be type of irrigation method.

(d) Nomimal: There is no ordering to the types of stimulus.

(e) Interval: The yield can be regarded as a point on the positive number line. The difference between two values provides a measure of the amount by which the yields differ.

(f) Ordinal: The ratings can be ordered from smallest to largest. However, the difference between 4 and 3 is not necessarily the same as the difference between 3 and 2.

(g) Nominal: The values distinguish between distinct categories. However, they cannot be ordered in any way.

4.3. (a) The target population can be described graphically as

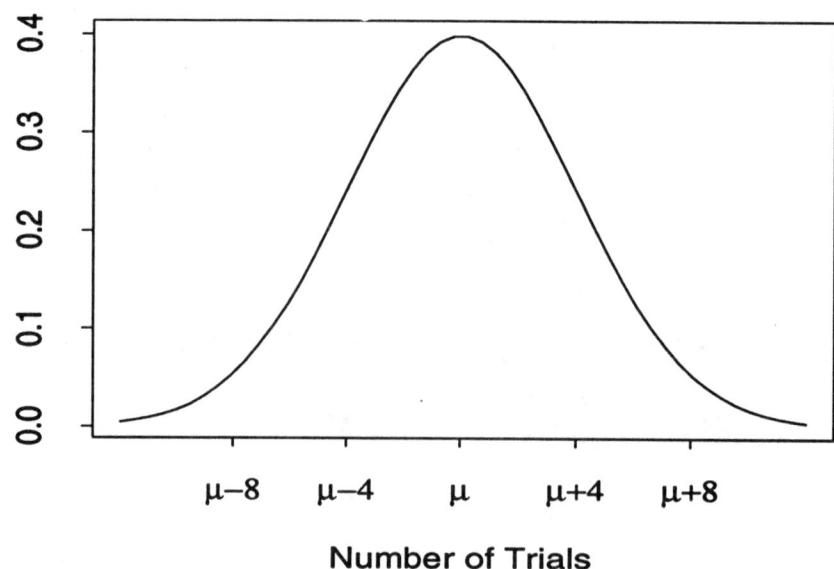

Number of Trials

(b) It means that in all experiments of this type $100\pi\%$ of the sample mean number of trials needed by the experimental rats will be between $\mu - 2$ and $\mu + 2$.

(c) To determine the value of π we first find the two z-scores corresponding to the values that are two units from the mean. Note that $\sigma_{\overline{Y}} = 4/\sqrt{10} = 1.2649$. The two z-scores are

$$z_1 = \frac{(\mu - 2) - \mu}{1.2649} = -1.5811$$

and

$$z_2 = \frac{(\mu + 2) - \mu}{1.2649} = +1.5811$$

From Table C.1 we find that the tail probabilities associated with these z-scores are 0.0571. Because π is the area under the normal curve between these two z-scores, the value of π is $1 - 2(0.0571) = 0.8858$.

That is, 88.58% of all sample mean number of trials needed by rats will be within two units of the population mean.

(d)   i. A 95% confidence interval for μ is

$$\hat{\theta} - z(\alpha/2)\sigma_{\hat{\theta}} \leq \theta \leq \hat{\theta} + z(\alpha/2)\sigma_{\hat{\theta}}$$

or

$$15.7 - 1.96(1.2649) = 13.2208 \quad \text{and} \quad 15.7 + 1.96(1.2649) = 18.1792.$$

With 95% confidence, the population mean number of trials needed by the experimental rats is between 13.2208 and 18.1792.

ii. **H$_0$:** $\mu \geq 18$

H$_1$: $\mu < 18$

Test Statistic: $z_c = -1.8183$ with a p-value of approximately 0.0344

Conclusion: With a p-value of 0.0344 there is enough evidence to say that, on the average, a rat needs less than 18 trials to learn to recognize the signal.

4.5. (a) We assume that the indices are a random sample from a normal population.

(b) The summary statistics for the data are

$$\overline{Y} = 60.6875 \qquad s^2 = 68.2291 \qquad s/\sqrt{n} = 8.2601/\sqrt{16} = 2.0650.$$

The endpoints of the interval are

$$\overline{Y} - t(15, 0.05)s/\sqrt{n} = 60.6875 - 1.7531(2.0650) = 57.0673$$

and

$$\overline{Y} + t(15, 0.05)s/\sqrt{n} = 60.6875 + 1.7531(2.0650) = 64.3077.$$

The 90% confidence interval for the mean of the population is $(57.0673, 64.3077)$.

(c) The psychologist can conclude with 90% confidence that the mean index of all children attending elementary schools in the particular school district is between 57.0673 and 64.3077.

(d) The test statistic is

$$t_c = \frac{\overline{Y} - \mu_0}{\hat{\sigma}_{\overline{Y}}} = \frac{60.6875 - 50}{2.0650} = 5.1755.$$

The p-value for this test is less than 0.001, because the t-statistic is larger than any value in Table C.2.

(e) The psychologist can conclude, at any α-level greater than or equal to 0.001, that the mean index is greater than 50.

(f) If the psychologist used the confidence interval approach, then the test can be performed at the 5% level because the lower end point of a 90% interval is a 95% lower confidence bound for μ. Because the 95% lower bound is greater than 50, the psychologist can conclude that the mean index is greater than 50.

4.7. (a) Let Population 1 be Urea + N-serve and Population 2 be Urea. Then summary statistics for the two samples are

i	n_i	$\overline{Y}_i$	s_i
1	15	11.1467	2.1397
2	13	8.5846	2.2883

This gives

$$\overline{Y}_1 - \overline{Y}_2 = 11.1467 - 8.5846 = 2.5621,$$

$$s_p^2 = \frac{(n_1 - 1)s_1^2 + (n_2 - 1)s_2^2}{N - 2} = \frac{14(4.5784) + 12(5.2368)}{26}$$

and

$$\hat{\sigma}_{\overline{Y}_1 - \overline{Y}_2} = \sqrt{4.8821\left(\frac{1}{15} + \frac{1}{13}\right)} = 0.8372.$$

With this information, the lower bound is

$$\overline{Y}_1 - \overline{Y}_2 - t(26, 0.05)\hat{\sigma}_{\overline{Y}_1 - \overline{Y}_2} = 2.5621 - 1.7506(0.8372) = 1.0965$$

(b) With 95% confidence, the true mean differential N-loss between two fertilizer regimen is at least 1.0965.

(c) Under the assumption that the unknown variances are not necessarily equal, the standard deviation of $\overline{Y}_1 - \overline{Y}_2$ becomes

$$\hat{\sigma}_{\overline{Y}_1 - \overline{Y}_2} = \sqrt{\frac{4.5784}{15} + \frac{5.2364}{13}} = 0.8414$$

The degrees of freedom, ν, is the largest integer less than or equal to k where

$$\frac{1}{k} = \frac{1}{n_1 - 1}\left[\frac{s_1^2/n_1}{s_1^2/n_1 + s_2^2/n_2}\right]^2 + \frac{1}{n_2 - 1}\left[\frac{s_2^2/n_2}{s_1^2/n_1 + s_2^2/n_2}\right]^2$$
$$= \frac{1}{14}\left[\frac{4.5784/15}{4.5784/15 + 5.2364/13}\right]^2 + \frac{1}{12}\left[\frac{5.2364/13}{4.5784/15 + 5.2364/13}\right]^2$$
$$= 0.0133 + 0.0269 = 0.0403.$$

Thus, $k = 1/0.0403 = 24.8326$ and $\nu = 24$.

Now, the lower bound is

$$(\overline{Y}_1 - \overline{Y}_2) - t(24, 0.05)\hat{\sigma}_{\overline{Y}_1 - \overline{Y}_2} = 2.5621 - 1.7109(0.8414) = 1.1225.$$

With 95% confidence, the true mean differential N-loss between two fertilizer regimen is at least 1.1225.

(d) The factor to be considered is the information known about the variances. If the variances are known, then the first lower bound of 1.32% can be used. If the variances are not known but can be considered equal, then the bound of 1.0965% can be used. Finally, if the variances are not known and can't be considered equal, then the bound of 1.1225% can be used.

4.9. (a) We assume that the observations are independent random samples from two normal populations with a common variance.

(b) Let the population of Holstein cows be Population 1 and the population of Guernsey cows be Population 2. Then summary statistics on the two samples are

i	n_i	$\overline{Y}_i$	s_i
1	13	6036.46	1515.42
2	10	5508.9	866.96

Then,
$$\hat{\theta} = \overline{Y}_1 - \overline{Y}_2 = 6036.46 - 5508.9 = 527.56,$$
$$s_p^2 = \frac{(n_1 - 1)s_1^2 + (n_2 - 1)s_2^2}{n_1 + n_2 - 2} = \frac{12(2296497) + 9(751619)}{21} = 1634406.$$
$$\hat{\sigma}_{\hat{\theta}} = \hat{\sigma}_{\overline{Y}_1 - \overline{Y}_2} = \sqrt{s_p^2\left(\frac{1}{n_1} + \frac{1}{n_2}\right)} = \sqrt{1634406\left(\frac{1}{13} + \frac{1}{10}\right)} = 537.7398.$$

Then, the 95% confidence lower bound is

$$\hat{\theta} - t(21, 0.05)\hat{\sigma}_{\hat{\theta}} = 527.56 - 1.7207(537.7398) = -397.7289.$$

(c) With 95% confidence, the differential mean lymphocyte count between Holstein cows and Guernsey cows is at least -340.12. In other words, the mean lymphocyte count for Holstein cows is at most 340.12 less than the mean lymphocyte count for Guernsey cows.

(d) The point estimate of the parameter, $\hat{\theta} = 527.56$, stays the same for this bound. The standard error now becomes

$$\hat{\sigma}_{\hat{\theta}} = \sqrt{\frac{s_1^2}{n_1} + \frac{s_2^2}{n_2}} = \sqrt{\frac{2296497}{13} + \frac{751619}{10}} = 501.81221.$$

The degrees of freedom, ν, for the confidence interval can be found via Satterthwaite's approximation. It is the largest integer less than or equal to k, where

$$1/k = \frac{1}{n_1 - 1}\left[\frac{s_1^2/n_1}{s_1^2/n_1 + s_2^2/n_2}\right]^2 + \frac{1}{n_2 - 1}\left[\frac{s_2^2/n_2}{s_1^2/n_1 + s_2^2/n_2}\right]^2$$

$$= \frac{1}{12}\left[\frac{2296497/13}{2296497/13 + 751619/10}\right]^2 + \frac{1}{9}\left[\frac{751619/10}{2296497/13 + 751619/10}\right]^2$$

$$= 0.0509.$$

This gives $k = 19.6464$ and $\nu = 19$. Therefore, the 95% confidence lower bound is

$$\hat{\theta} - t(19, 0.05)\hat{\sigma}_{\hat{\theta}} = 527.56 - 1.7291(501.8122) = -340.12.$$

(e) The two sample variances are $s_1^2 = 2296497$ and $s_2^2 = 751619$. By calculating $s_1^2/s_2^2 = 3.055$ we see that the variance for the Holstein cows is three times the variance for the Guernsey cows. Therefore, the interval which does not assume equal variances might be more appropriate.

4.11. (a) The measurements before and after infusion of the drug are paired because they were taken on the same lamb. The paired t-test takes advantage of this relationship. To perform this method we first calculate the differences between each pair.

Lamb	PVR before histamine	PVR after histamine	Difference
1	0.095	0.176	0.081
2	0.106	0.142	0.036
3	0.082	0.194	0.112
4	0.152	0.136	−0.016
5	0.090	0.115	0.025
6	0.086	0.084	−0.002
7	0.137	0.103	−0.034
8	0.121	0.189	0.068

The null hypothesis is H_0: $\mu_D \leq 0$ or the mean PVR after histamine is greater than the mean PVR before histamine. The alternative hypothesis is H_0: $\mu_D > 0$ or the mean PVR after histamine is less than the mean PVR before histamine.

The test statistic is

$$s_D = 0.0506$$

$$\overline{D} = 0.0337$$

$$t_c = \frac{\overline{D} - 0}{s_D/\sqrt{n}} = \frac{0.0337}{0.0506/\sqrt{8}} = 1.8837.$$

The 0.10-level critical value for this test is $t(7, 0.10) = 1.4149$. Using the test statistic method we reject H_0 at level 0.10 if

$$t_c > t(7, 0.10).$$

Because $1.8837 > 1.4149$ we conclude at the 0.10 level that the data shows that histamine does increase PVR on the average.

(b) The 95% confidence interval is

$$\left(\overline{D} - t(7, 0.025)\hat{\sigma}_{\overline{D}}, \overline{D} + t(7, 0.025)\hat{\sigma}_{\overline{D}}\right).$$

The endpoints are

$$\overline{D} - t(7, 0.025)\hat{\sigma}_{\overline{D}} = -0.0337 - 2.3646\left(0.0506/\sqrt{8}\right) = -0.0760$$

and

$$\overline{D} + t(7, 0.025)\hat{\sigma}_{\overline{D}} = -0.0337 + 2.3646\left(0.0506/\sqrt{8}\right) = 0.0086.$$

Hence, the 95% confidence interval is

$$-0.0760 \leq \mu_D \leq 0.0086.$$

These bounds are calculated with 95% confidence because if a large number of paired samples of size eight were taken and an interval were calculated for each sample using the same procedure, then 95% of the intervals would contain the true value for μ_D.

4.13. (a) The sample standard deviation estimates the population standard deviation. That is the standard deviation of the population of numbers of trials needed to learn to recognize the signal.

(b) In Exercise **4.3** we have $n = 10$ and $s^2 = 9.7888$. The 95% confidence interval for the population variance is

$$\left\{\frac{(n-1)s^2}{\chi^2(n-1, 0.025)}, \frac{(n-1)s^2}{\chi^2(n-1, 1-0.025)}\right\} = \left\{\frac{9(9.7888)}{19.0228}, \frac{9(9.7888)}{2.7004}\right\}$$

which gives $(4.6312, 32.6244)$. Taking the square root of these endpoints gives the 95% confidence interval for the population standard deviation, i.e. $(2.1520, 5.7118)$.

(c) With 95% confidence, the standard deviation of the population of number of trials is between 2.1520 and 5.7118.

(d) We assume that the numbers of trials are a random sample from a normal population.

4.15. (a) To construct a 95% confidence interval for σ_B/σ_G we start with

$$s_G = 2.4374 \quad s_G^2 = 5.9409 \quad n_G = 11$$
$$s_B = 2.0854 \quad s_B^2 = 4.3489 \quad n_B = 16$$

Then a 95% confidence interval for σ_B^2/σ_G^2 is

$$\frac{s_B^2}{s_G^2}\frac{1}{F(15, 10, 0.025)} \leq \frac{\sigma_B^2}{\sigma_G^2} \leq \frac{s_B^2}{s_G^2}\frac{1}{F(15, 10, 0.975)}$$

where

$$F(15, 10, 0.975) = \frac{1}{F(10, 15, 0.025)} = \frac{1}{3.060} = 0.3268.$$

So that we have

$$\frac{4.3489}{5.9409}\frac{1}{3.522} \leq \frac{\sigma_B^2}{\sigma_G^2} \leq \frac{4.3489}{5.9409}\frac{1}{0.3268}$$
$$0.2078 \leq \frac{\sigma_B^2}{\sigma_G^2} \leq 2.2400.$$

Now, to get a 95% confidence interval for σ_B/σ_G we take the square root of the endpoints, i.e.

$$0.4558 \leq \frac{\sigma_B}{\sigma_G} \leq 1.4967.$$

(b) With 95% confidence we can conclude that the standard deviation for boys is at least 46% and at most 150% of the standard deviation for girls.

(c) We assume that the dental measurements of the eleven girls and the sixteen boys are independent random samples from two normal populations.

4.17. From S-plus, the confidence interval and hypothesis test are given in the following output.

```
> t.test(hcl,conf=0.90)

        One-sample t-Test

data:  hcl
t = 32.9382, df = 34, p-value = 0
alternative hypothesis: true mean is not equal to 0
90 percent confidence interval:
 1.857755 2.058817
sample estimates:
 mean of x
  1.958286
```

4.19. The hypothesis test and 95% confidence interval are given in the following S-plus output.

```
> t.test(pvr$before,pvr$after,paired=T)

        Paired t-Test

data:  pvr$before and pvr$after
t = -1.8849, df = 7, p-value = 0.1014
alternative hypothesis: true mean of differences is not equal to 0
95 percent confidence interval:
 -0.076089204  0.008589204
sample estimates:
 mean of x - y
      -0.03375
```

4.21. Let Y_j denote the length of the jth ear. Then the model can be written as

$$Y_j = \mu + E_j \qquad j = 1, 2, 3, \ldots, 15$$

where

μ = the population mean length of ears in plants of this hybrid variety,

E_j = the random error in Y_j; the E_j are independent $N(0, \sigma^2)$.

4.23. Let Y_{ij} denote the ith measurement in the jth group. Then

$$Y_{ij} = \mu_i + E_{ij} \qquad i = 1, 2; j = 1, 2, \ldots, 11 \text{ if } i = 1 \text{ and } j = 1, 2, \ldots, 16 \text{ if } i = 2.$$

where

μ_i = population mean dental measurement in group i,

E_{ij} = random error in Y_{ij}; the E_{ij} are independently distributed as $N(0, \sigma^2)$.

4.25. (a) From Exercise **4.4**, $\overline{Y} = 1.96$, $s = 0.3517$, and $n = 35$. Also, $t(34, 0.05) \approx t(30, 0.05) = 1.6973$. Then, the 95% lower prediction bound is

$$\overline{Y} - t(34, 0.05)\sqrt{\left(\frac{1}{5} + \frac{1}{35}\right)(0.3517)^2} = 1.96 - 1.6973(0.1681) = 1.6747.$$

(b) With 95% confidence the average infection rate of a sample of five HCL patients is at least 1.6747.

(c) The population mean infection rate is a fixed value while the mean infection rate of a sample of five HCL patients is a random value. The 95% lower prediction bound for the mean infection rate of a sample of five HCL patients must be smaller, i.e. the interval is wider, in order to account for the extra variability that the random variable has but the population mean does not.

4.27. (a) 95% normal limits are the 95% prediction limits. These are

$$\left(\overline{Y} - t(999, 0.025)\sqrt{\left(1 + \frac{1}{n}\right)s^2} , \quad \overline{Y} + t(999, 0.025)\sqrt{\left(1 + \frac{1}{n}\right)s^2} \right) =$$

$$\left(1.1 - 1.9719\sqrt{\left(1 + \frac{1}{1000}\right)(0.35)^2} , \quad 1.1 + 1.9719\sqrt{\left(1 + \frac{1}{1000}\right)(0.35)^2} \right) =$$

$$(0.4064, \quad 1.7936)$$

(b) Only 5% of the disease-free population have creatinine level outside the range $(0.41, 1.79)$. Only 2.5% of the disease-free population will have levels above 1.79.

(c) In view of our answer in part (b), a subject with creatinine level of 2.00 is likely to be a subject with renal damage.

4.29. The following figure shows the boxplot of the ear length data

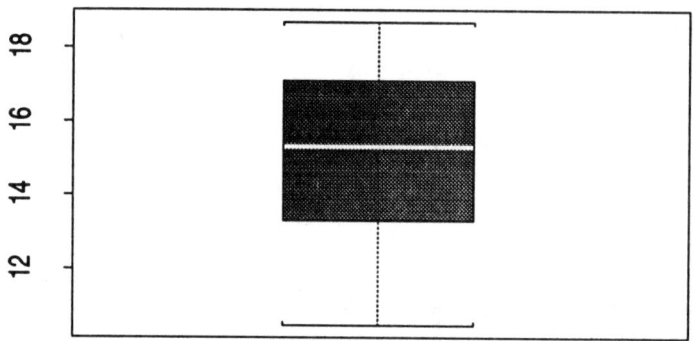

Ear Length

(a) The boxplot is symmetric enough about the median line to consider it not skewed. The lower whisker is a little longer than the upper one, but that is not unexpected with small sample sizes.

(b) The boxplot does not show any points outside of the whiskers. Any points outside of the whiskers would be considered extreme values. Thus, there is no indication that the population has heavy tails.

4.31. The following figure shows the boxplots of the data in Exercise **4.12.**

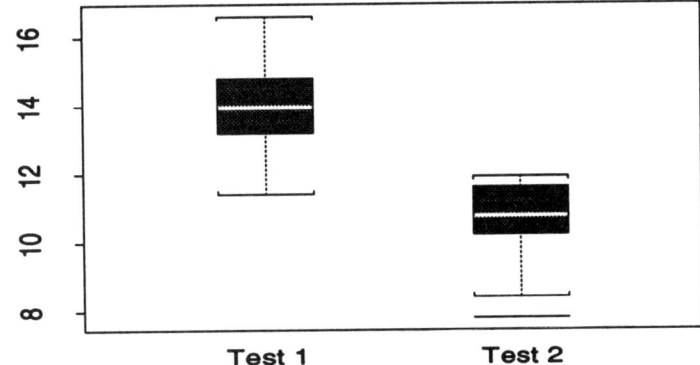

(a) The distribution for test 1 is symmetric and for test 2 is skewed towards lower values.

(b) Both distributions have heavy tails. The whiskers of the boxplots are as long as the box. I.e., the tails are as wide as the middle 50% of the data.

Chapter 5

Inferences About One or Two Populations: Ordinal Data

5.1. (a) The groups are

$$\{L, L, L, L\} \qquad \{M, M, M\} \qquad \{H\}.$$

(b) Without ties the ranks would be $1, 2, 3, \ldots, 8$. With ties the ranks are

Subject	SES	Rank
1	L	$(1+2+3+4)/4 = 2.5$
2	L	2.5
3	H	8
4	M	$(5+6+7)/3 = 6$
5	L	2.5
6	M	6
7	L	2.5
8	M	6

5.3. (a) After combining the data and sorting the measurements from smallest to largest we have

Breed	H	H	G	G	G	H	H	H
Count	3800	4214	4273	4497	4600	4900	5002	5135
Rank	1	2	3	4	5	6	7	8
Breed	H	G	G	G	H	H	G	G
Count	5166	5182	5407	5509	6080	6205	6295	6310
Rank	9	10	11	12	13	14	15	16
Breed	G	G	H	H	H	H	H	
Count	6425	6591	6700	7031	7290	8043	8908	
Rank	17	18	19	20	21	22	23	

Since there are no ties, the tie sizes are all one and the tie groups are the measurements themselves.

(b) Since there are no ties, the ranks are the numbers one through twenty-three.

(c) The ranks for the Holstein cows are

$$1, 2, 6, 7, 8, 9, 13, 14, 19, 20, 21, 22, 23.$$

The mean rank is 12.6923. The ranks for the Guernsey cows are

$$3, 4, 5, 10, 11, 12, 15, 16, 17, 18.$$

The mean rank is 11.1.

Because the mean rank for the Holstein cows is larger than the mean rank for the Guernsey cows, we can conclude that the Holstein cows as a group had a higher lymphocyte count than the group of Guernsey cows.

5.5. (a) We want to test $H_0: \eta \geq 13$ vs $H_1: \eta < 13$. The sorted values from the fertilizer UN samples are

$$8.0 \quad 8.7 \quad 9.0 \quad 9.5 \quad 9.8 \quad 10.0 \quad 10.3 \quad 10.5$$
$$10.8 \quad 11.8 \quad 12.8 \quad 13.5 \quad 13.8 \quad 14.0 \quad 14.7$$

Hence, there are four values greater than 13 and $B_c = 4$. There are no zero differences in the modified sample, $\{Y_1 - 13, Y_2 - 13, Y_3 - 13, \ldots, Y_{15} - 13\}$, so $n^* = 15$.

There is not a $b(15, 0.05)$ value in Table C.5. The closest value is $b(15, 0.059) = 11$. Using this value, the rejection criteria for the 0.059−level test is

$$\text{Reject } H_0: \eta \geq 13 \text{ if } B_c \leq 15 - b(n^*, 0.059) = 15 - 11 = 4.$$

Because $B_c = 4$, we reject H_0 and conclude at the 0.059 level that the true median N-loss is less than 13%.

(b) The test statistic is $B_c = 4$. Looking up $n^* - B_c = 15 - 4 = 11$ in Table C.5 gives a p−value $= 0.059$. We could conclude at any α−level up to 0.059 that the true median N-loss is less than 13%.

(c) The t-test would be

$H_0: \mu \geq 13$

$H_1: \mu < 13$

Test Statistic: $t_c = \dfrac{\overline{Y} - \mu}{s/\sqrt{n}} = \dfrac{11.1467 - 13}{2.1397/\sqrt{15}} = -3.3546$

Rejection Region: Reject H_0 if $t_c \leq -t(14, 0.05) = -1.7613$

Conclusion: Because $t_c = -3.3546$, we can conclude at the 0.05 significance level that the true mean N-loss is less than 13%.

From Table C.2 we find $0.005 < p$−value < 0.001. This p−value is much smaller than the p−value for the sign test.

The sign test would be chosen over the t-test if there were serious violations in the assumption of normality. If not, then the t-test could be chosen.

5.7. (a) The model is

$$Y_j = \eta + E_j \qquad j = 1, 2, \ldots, n$$

where

$Y_j =$ the jth differential CPZ value,

$\eta =$ the population median, and

$E_j = Y_j - \eta$ is the error term in the jth value; the E_j are a random sample from a continuous symmetric population with zero median.

(b) We want to test $H_0: \eta_D \geq 1.5$ versus $H_1: \eta_D < 1.5$. First, we calculate the differences of Hanks' solution − Sucrose. These are $-4, 2, 0, 2, 1, -8, -7, 1$. Then the hypothesized value, 1.5, is subtracted from each of these.

original	−4	−2	0	2	1	−8	−7	1
non−zero values	−5.5	0.5	−1.5	0.5	−0.5	−9.5	−8.5	−0.5
absolute values	5.5	0.5	1.5	0.5	0.5	9.5	8.5	0.5
absolute ranks	6	2.5	5	2.5	2.5	8	7	2.5
signed ranks	0	2.5	0	2.5	0	0	0	0

Then, adding up the signed ranks gives $W_c^+ = 5.0$. Next, $n = 8$, $N = \frac{n(n+1)}{2} = \frac{8(9)}{2} = 36$, and $w(8, 0.055) = 30$. Reject H_0 at level 0.055 if $W_c^+ \leq N - w(n, \alpha) = 36 - 30 = 6$. Since, $5 < 6$ we reject H_0 and conclude that the population of differences is distributed with median less than 1.5.

(c) $N - w(n, 0.055) = 36 - 30 = 6$ and $N - w(n, 0.027) = 36 - 32 = 4$. Since $W_c^+ = 5$ is between these two values, we can say that the $p-$value is between 0.027 and 0.055.

5.9. (a) To calculate the confidence interval we first calculate the Walsh averages. The order statistics and the Walsh averages are

	0.4	1.2	1.3	1.3	1.4	2.1	2.1	2.2	2.4	2.8
0.4	0.4	0.8	0.85	0.85	0.9	1.25	1.25	1.3	1.4	1.6
1.2		1.2	1.25	1.25	1.3	1.65	1.65	1.7	1.8	2
1.3			1.3	1.3	1.35	1.7	1.7	1.75	1.85	2.05
1.3				1.3	1.35	1.7	1.7	1.75	1.85	2.05
1.4					1.4	1.75	1.75	1.8	1.9	2.1
2.1						2.1	2.1	2.15	2.25	2.45
2.1							2.1	2.15	2.25	2.45
2.2								2.2	2.3	2.5
2.4									2.4	2.6
2.8										2.8
3										
⋮										
12.5										

	3.0	4.5	4.6	4.6	4.8	5.7	8.0	8.8	11.4	12.5
⋮	⋮	⋮	⋮	⋮	⋮	⋮	⋮	⋮	⋮	⋮
0.4	1.7	2.45	2.5	2.5	2.6	3.05	4.2	4.6	5.9	6.45
1.2	2.1	2.85	2.9	2.9	3	3.45	4.6	5	6.3	6.85
1.3	2.15	2.9	2.95	2.95	3.05	3.5	4.65	5.05	6.35	6.9
1.3	2.15	2.9	2.95	2.95	3.05	3.5	4.65	5.05	6.35	6.9
1.4	2.2	2.95	3	3	3.1	3.55	4.7	5.1	6.4	6.95
2.1	2.55	3.3	3.35	3.35	3.45	3.9	5.05	5.45	6.75	7.3
2.1	2.55	3.3	3.35	3.35	3.45	3.9	5.05	5.45	6.75	7.3
2.2	2.6	3.35	3.4	3.4	3.5	3.95	5.1	5.5	6.8	7.35
2.4	2.7	3.45	3.5	3.5	3.6	4.05	5.2	5.6	6.9	7.45
2.8	2.9	3.65	3.7	3.7	3.8	4.25	5.4	5.8	7.1	7.65
3	3	3.75	3.8	3.8	3.9	4.35	5.5	5.9	7.2	7.75
4.5		4.5	4.55	4.55	4.65	5.1	6.25	6.65	7.95	8.5
4.6			4.6	4.6	4.7	5.15	6.3	6.7	8	8.55
4.6				4.6	4.7	5.15	6.3	6.7	8	8.55
4.8					4.8	5.25	6.4	6.8	8.1	8.65
5.7						5.7	6.85	7.25	8.55	9.1
8							8	8.4	9.7	10.25
8.8								8.8	10.1	10.65
11.4									11.4	11.95
12.5										12.5

There are $n = 20$ values which is outside of the range of Table C.7. To use the large-sample approximation the tie groups are first calculated. These are, with their sizes,

Group	0.4	1.2	1.3	1.4	2.1	2.2	2.4	2.8	3
Size	1	1	2	1	2	1	1	1	1

Group	4.5	4.6	4.8	5.7	8	8.8	11.4	12.5
Size	1	2	1	1	1	1	1	1

Now, the mean and standard deviation of W^+ are

$$\mu_{W^+} = \frac{N}{2} = \frac{20(21)/2}{2} = \frac{210}{2} = 105$$

and

$$\sigma_{W^+} = \sqrt{\frac{1}{12}\left[N(2n+1) - \frac{1}{4}\sum_{j=1}^{17} t_j(t_j^2 - 1)\right]}$$

$$= \sqrt{\frac{1}{12}\left[210(2(20)+1) - \frac{1}{4}18\right]} = \sqrt{717.125}$$

$$= 26.7792$$

Then an approximation to the 95% critical value is

$$w(20, 0.10/2) \approx \mu_{W^+} + z(0.10/2)\sigma_{W^+} = 105 + 1.645(26.7792) = 149.0518 \approx 150.$$

So that $U = w(20, 0.05) \approx 150$ and $L = N + 1 - w(20, 0.05) \approx 210 + 1 - 150 = 61$. Hence, the 90% confidence interval is

$$W_{(61)} < \eta < W_{(150)}$$

or, from the sorted Walsh averages,

$$2.5 < \eta < 5.4.$$

Therefore, with 90% confidence the true median time elapsed before the redness in the irritated eye returned to the color of the un-irritated eye is between 2.5 minutes and 5.4 minutes.

(b) From the histogram and box-plots, it can be seen that the data is skewed to the right and there are some large extreme values. Therefore, the assumption that the population is symmetric may be unreasonable for these data.

5.11. (a) The model can be written as

$$Y_{ij} = \eta_i + E_{ij} \qquad j = 1, 2, \ldots, n_i; \; i = 1, 2.$$

where

$Y_{ij} =$ the jth dental measurement in the ith group,

$\eta_i =$ mean dental measurement of the ith population,

$E_{ij} =$ the random error in Y_{ij}; the E_{ij} are independent random samples from a continuous population with mean zero.

(b) Let the population of dental measurements of boys be population 2. Then the null and alternative hypotheses are

$$H_0 : \theta = \eta_2 - \eta_1 \le 0 \quad \text{vs} \quad H_1 : \theta = \eta_2 - \eta_1 > 0.$$

To calculate the test statistic we first determine the ranks of the values in the combined samples. The sorted values with their ranks are

Gender	Girls	Girls	Girls	Girls	Girls	Girls	Girls	Boys	Boys
Msmt	19.5	21.5	22.5	23.0	23.5	24.0	25.0	25.0	25.0
Rank	1	2	3	4	5	6	8	8	8

Gender	Girls	Boys	Girls	Boys	Boys	Boys	Girls	Boys	Boys
Msmt	25.5	25.5	26.0	26.0	26.0	26.0	26.5	26.5	26.5
Rank	10.5	10.5	13.5	13.5	13.5	13.5	17.0	17.0	17.0

Gender	Boys	Boys	Girls	Boys	Boys	Boys	Boys	Boys	Boys
Msmt	27.0	27.5	28.0	28.0	28.5	29.5	30.0	31.0	31.5
Rank	19.0	20.0	21.5	21.5	23.0	24.0	25.0	26.0	27.0

Now, by summing the ranks of the sample of boys, we get the test statistic.

$$W_c = R_1 + R_2 + \cdots + R_{16} = 8 + 8 + \cdots + 27.0 = 286.5.$$

Because $n_1 = 11$ and $n_2 = 16$ are outside the range of Table C.7, the large sample approximation will be used. The mean and standard deviation of W are

$$\mu_W = \frac{n_2(N+1)}{2} = \frac{16(28)}{2} = 224$$

and

$$\sigma_W = \sqrt{\frac{n_1 n_2}{12}\left[N + 1 - \frac{\sum_{j=1}^{g} t_j(t_j^2 - 1)}{N(N+1)}\right]}.$$

There are eighteen tied groups and the t_j are

j	1	2	3	4	5	6	7	8	9
t_j	1	1	1	1	1	1	3	2	4
Rank	1	2	3	4	5	6	8	10.5	13.5

j	10	11	12	13	14	15	16	17	18
t_j	3	1	1	2	1	1	1	1	1
Rank	17	19	20	21.5	23	24	25	26	27

This gives

$$\sigma_W = \sqrt{\frac{11(16)}{12}\left[27 + 1 - \frac{120}{27(28)}\right]} = 20.2074.$$

Now,

$$w(11, 16, 0.05) \approx \mu_W + z(0.05)\sigma_W = 224 + 1.645(20.2074) = 257.2412.$$

Reject H_0: $\theta \leq 0$ if $W_c \geq w(11, 16, 0.05) \approx 257.2412$. Hence, at the 0.05 significance level we conclude that the data does show that the mean dental measurement for boys is higher than that for girls.

(c) An approximate 95% confidence interval for the difference between the mean dental measurements of the boys and girls is

$$D_{(L)} \leq \theta \leq D_{(U)}$$

where

$$U = w(n_1, n_2, \alpha/2) - \frac{n_2(n_2 + 1)}{2}$$
$$L = \frac{n_2(N + n_1 + 1)}{2} + 1 - w(n_1, n_2, \alpha/2).$$

First, the $n_1 n_2 = 11(16) = 176$ differences, $D_{ij} = Y_{2j} - Y_{1i}$, are calculated. Because there are so many they will not be shown here. In fact, it is best to use computer software to do this exercise.

Because $n_1 = 11$ and $n_2 = 16$ are outside the range of Table C.7, the large sample approximation will be used, i.e.

$$w(11, 16, 0.025) \approx \mu_W + z(0.025)\sigma_W \approx 224 + 1.96(20.2074) \approx 263.$$

Next, $U = 263 - 16(17)/2 = 126$ and $L = 16(27 + 11 + 1)/2 + 1 - 263 = 50$.

Finally, from the sorted values of D_{ij}, $D_{(50)} = 1.5$ and $D_{(127)} = 5.0$. Therefore, with approximately 95% confidence the difference between the mean dental measurements of the boys and girls is between 1.5mm and 5.0mm. Or with approximately 95% confidence the mean dental measurement for boys is as much as 5.0mm or as little as 1.5mm larger than the mean dental measurement for girls.

(d) The 95% confidence interval from Exercise **4.8** is $(-5.179, -1.5768)$. (Note that this is for $\mu_1 - \mu_2$ and the confidence interval in Part **c** in Exercise **5.11** is for $\eta_2 - \eta_1$.) Other than the sign change the confidence intervals are very similar. Thus, the confidence interval is valid regardless of whether or not the assumption of normal disbribution is satisfied.

5.13. (a) The new values are

Group	serum albumin levels			
Marasmus	3.0	3.2	3.6	
Control	5.1	4.7	25.1	4.8

After $\theta_0 = 1.5$ is subtracted from the Control values, the new ranks are

Group	M	M	C	C	M	C	C
Value	3.0	3.2	3.2	3.3	3.6	3.6	23.6
Rank	1	2.5	2.5	4	5.5	5.5	7

Notice that the ranks have changed only a little. Now

$$W_c = 2.5 + 4 + 5.5 + 7 = 19.$$

The critical value is the same, $w(3, 4, 0.029) = 18$, because the sample sizes are the same. The conclusion is even the same: We conclude at the 0.029-level that the median serum albumin level for the control children is 1.5g/dliter more than the median level for children with the Marasmus form of PEM.

(b) To do the test, we first calculate the summary statistics.

Group	1 (Marasmus)	2 (Control)
n_i	3	4
$\overline{Y}_i$	3.2667	9.925
s_i	0.3055	10.1181
s_i^2	0.0933	102.3759

Then

$$\hat{\sigma}_{\hat{\theta}} = \sqrt{\frac{0.0933}{3} + \frac{102.3759}{4}} = 5.0621.$$

The degrees of freedom ν is the largest integer less than k, where

$$\frac{1}{k} = \frac{1}{n_1 - 1}\left[\frac{s_1^2/n_1}{s_1^2/n_1 + s_2^2/n_2}\right]^2 + \frac{1}{n_2 - 1}\left[\frac{s_2^2/n_2}{s_1^2/n_1 + s_2^2/n_2}\right]^2$$

$$= \frac{1}{2}\left[\frac{0.0933/3}{0.0933/3 + 102.3759/4}\right]^2 + \frac{1}{3}\left[\frac{102.3759/4}{0.0933/3 + 102.3759/4}\right]^2$$

$$= 0.3325.$$

Thus, ν is the largest integer less than 3.0073, i.e. $\nu = 3$. Then, the hypothesis test is

H$_0$: $\mu_1 - \mu_2 \leq 1.5$

H$_1$: $\mu_1 - \mu_2 > 1.5$

Test Statistic: $t_c = \dfrac{\overline{Y}_1 - \overline{Y}_2}{\hat{\sigma}_{\hat{\theta}}} = \dfrac{3.2667 - 9.925}{5.0621} = -1.3153$

Rejection Region: Reject H$_0$ if $t_c < -t(3, 0.05) = -2.3534$.

Conclusion: We conclude at the 0.05 significance level that the data does not show that the mean serum level for control children is 1.5 g/dliter more than the median level for children with Marasmus form of PEM.

(c) The Wilcoxon rank sum test, like other procedures based on ranks, is not sensitive to presence of outliers in the data.

Chapter 6

Inferences About One or Two Populations: Categorical Data

6.1. (a) $C = 2$; π_1 is the proportion of exposed homes in the population and π_2 is the proportion of non-exposed homes in the population.

Exposed	0	1
Proportion	π_1	π_2

(b) $C = 4$.

Litter Size	1	2	3	4
Proportion	0.2	0.3	0.3	0.2

(c) $C = 8$.

No. of Leaves	15	16	17	18	19	20	21	22
Proportion	0.02	0.03	0.15	0.30	0.30	0.15	0.03	0.02

(d) $C = 2$.

Germinate	0	1
Proportion	0.20	0.80

(e) $C = 9$.

Age / Blood Pressure	Young Low	Young Normal	Young High	Middle Age Low	Middle Age Normal
Proportion	0.01	0.16	0.03	0.06	0.32

Age / Blood Pressure	Middle Age High	Old Low	Old Normal	Old High
Proportion	0.08	0.04	0.16	0.14

(f) $C = 12$.

Errors in Tests	(0,0)	(0,1)	(0,2)	(0,3)	(1,0)	(1,1)
Proportion	0.04	0.03	0.06	0.04	0.09	0.24

Errors in Tests	(1,2)	(1,3)	(2,0)	(2,1)	(2,2)	(2,3)
Proportion	0.14	0.08	0.04	0.06	0.14	0.04

(g) $C = 7$.

No. of Caterpillars	10	11	12	13	14	15	16
Proportion	0.15	0.15	0.10	0.10	0.15	0.15	0.20

6.3. (a) Let π be the probability that a patient treated with the antibiotic will be cured of the infection. Then a 90% confidence interval for π is

$$\hat{\pi} \pm z(\alpha/2)\hat{\sigma}_{\hat{\pi}}$$

where $\hat{\pi} = 86/100 = 0.86$, and

$$\hat{\sigma}_{\hat{\pi}} = \sqrt{\frac{\hat{\pi}(1-\hat{\pi})}{n}} = \sqrt{\frac{0.86(1-0.86)}{100}} = 0.0350,$$

and $z(\alpha/2) = z(0.10/2) = z(0.05) = 1.645$.
Therefore, the interval is

$$0.86 \pm 1.645(0.0350) \rightarrow (0.8024, 0.9176).$$

With 90% confidence, the probability that a patient treated with the antibiotic will be cured of the infection is between 0.8024 and 0.9176.

(b) **H$_0$:** $\pi \geq 0.85$
H$_1$: $\pi < 0.85$

Test Statistic: $z_c = \dfrac{\hat{\pi} - \pi_0}{\hat{\sigma}_{\hat{\pi}}} = \dfrac{0.86 - 0.85}{0.0350} = 0.2857$
The $p-$value is $Pr\{z \leq 0.2857\} \approx 0.5 + 0.3897 = 0.8897$.

Conclusion: Because the $p-$value is so large the data does not support the conclusion that less than 85% of the patients will be cured by the new antibiotic.

6.5. (a) As defined in Exercise **6.4**, sensitivity is the probability that the test will give a positive result in a subject with the disease. Let Y be the number of positive results out of fifteen trials. Then Y can be regarded as a binomial random variable with probability π (which is the sensitivity) and $n = 15$. From the data the estimate of sensitivity is $\hat{\pi} = 13/15 = 0.8667$. That is, the sensitivity of the test is 86.67% or for 86.67% of the patients who actually have the disease, the test will correctly give a positive result.

(b) Because of the small sample size, Table C.8 will be used. We have $n = 15$, $f = 13$, and $1 - \alpha = 0.95$. However, $f = 13$ is not in the table, so $n - f = 15 - 13 = 2$ will be used. This gives $(0.0166, 0.4046)$ which is a confidence interval for $1 - \pi$. Subtracting this interval from 1 gives the 95% confidence interval for π or $(0.5954, 0.9834)$. With 95% confidence the sensitivity of the diagnostic test is between 0.5954 and 0.9834.

(c) The null and alternative hypotheses are H$_0$: $\pi \leq 0.80$ and H$_1$: > 0.80. The $p-$value is the probability of being greater than or equal to the observed value, i.e.

$$\begin{aligned}
p &= f(13|15, 0.80) + f(14|15, 0.80) + f(15|15, 0.80) \\
&= \frac{15!}{13!2!}(0.80)^{13}(0.20)^2 + \frac{15!}{14!1!}(0.80)^{14}(0.20)^1 + \frac{15!}{15!0!}(0.80)^{15}(0.20)^0 \\
&= 0.2309 + 0.1319 + 0.0352 \\
&= 0.398.
\end{aligned}$$

6.7. (a) A 99% confidence interval for π_1 can be gotten from Table C.8. The text gives $n = 15$, $1 - \alpha = 0.99$ and $f = 11$, but the latter is outside of the range of the table. Using $n - f = 15 - 11 = 4$ gives $(0.0488, 0.6273)$ as an interval for $1 - \pi$. A 99% confidence interval for π is $(0.3727, 0.9512)$. With 99% confidence, in a given sitting Judge 1 will rate between 37.27% and 95.12% of the wine as acceptable.

(b) The hypotheses are H$_0$: $\pi \leq 0.60$ and H$_1$: $\pi > 0.60$. Then, the $p-$value is

$$p = f(11|15, 0.60) + f(12|15, 0.60) + f(13|15, 0.60) + \cdots + f(15|15, 0.60) = 0.217.$$

Because the $p-$value is so large it is not reasonable to conclude that there is more than 60% chance that Judge 1 will find a sample of the wine acceptable.

(c) Table C.8 will be used because the sample size is only fifteen. The text gives $f = 8$ which is outside of the range of the table. $n - f = 15 - 8 = 7$ and $1 - \alpha = 0.95$ give an interval of $(0.2127, 0.7341)$ for $1 - \pi$. The interval for π is $(0.2659, 0.7873)$. With 95% confidence, in a given sitting both judges will rate between 26.59% and 78.73% of the wine as acceptable.

6.9. (a) Let π be the proportion of peas that are round and yellow. Then, the data estimates π as $\hat{\pi} = 110/200 = 0.55$ and

$$\hat{\sigma}_{\hat{\pi}} = \sqrt{\frac{0.55(0.45)}{200}} = 0.0352.$$

A 95% confidence interval for π is

$$\hat{\pi} \pm z(0.05/2)\hat{\sigma}_{\hat{\pi}} = 0.55 \pm 1.96(0.0352)$$

$= (0.4810, 0.6190)$. The population proportion of peas that are round and yellow is between 0.4810 and 0.6910.

(b) A $\chi^2()$ test will be conducted to determine whether the given data contradict the Mendelian theory. The expected frequencies are

$$\hat{f}_1 = f_+\pi_{10} = 200(9/16) = 112.5 \qquad \hat{f}_2 = f_+\pi_{20} = 200(3/16) = 37.5$$
$$\hat{f}_2 = f_+\pi_{30} = 200(3/16) = 37.5 \qquad \hat{f}_4 = f_+\pi_{40} = 200(1/16) = 12.5.$$

With these, then, the test statistic is

$$\chi^2(c) = \sum_{k=1}^{C} \frac{(f_k - \hat{f}_k)^2}{\hat{f}_k}$$
$$= \frac{(110 - 112.5)^2}{112.5} + \frac{(40 - 37.5)^2}{37.5} + \frac{(42 - 37.5)^2}{37.5} + \frac{(8 - 12.5)^2}{12.5} = 2.3823.$$

Now, the hypothesis test is

H$_0$: $\pi_1 = 9/16; \pi_2 = 3/16; \pi_3 = 3/16; \pi_4 = 1/16$
H$_1$: That the population proportions are not as specified by H$_0$.
Test Statistic: $\chi^2(c) = 2.3823$
Comparing this to a $\chi^2(3)$ distribution gives a $p-$value > 0.20.

Conclusion: With such a large $p-$value, there is not sufficient evidence to say that the population proportions are not as specified in the null hypothesis, i.e. as given by the Mendelian theory.

6.11. (a) The mean of this sample is

$$\hat{\lambda} = \overline{Y} = \frac{5(3) + 6(5) + 7(6) + \cdots + 13(2)}{50} = 8.80.$$

(b) To estimate the expected frequency for each category we first estimate the probabilities of an occurrence in each category. Let

$$\hat{f}(y) = \frac{e^{-\hat{\lambda}}(\hat{\lambda})^y}{y!}$$

be the estimated probability that exactly y revertants occur. Then, for category 1, the estimated probability of an occurrence is

$$Pr\{\text{occurrence in cat. 1}\} = \hat{f}(0) + \hat{f}(1) + \hat{f}(2) + \hat{f}(3) + \hat{f}(4) + \hat{f}(5)$$
$$= \frac{e^{-8.8}(8.8)^0}{0!} + \frac{e^{-8.8}(8.8)^1}{1!} + \cdots + \frac{e^{-8.8}(8.8)^5}{5!} = 0.1284$$

The estimated frequency in category 1 out of 50 assays is $50(0.1284) = 6.42$. The estimated probability of 6 revertants is

$$\hat{f}(6) = \frac{e^{-8.8}(8.8)^6}{6!} = 0.0972$$

and the estimated expected frequency is $50(0.0972) = 4.86$. Similarly, the estimated expected frequency for categories 3 through 8 are

Category	Probability	Est. Exp. Freq.
3	0.1222	6.11
4	0.1344	6.72
5	0.1314	6.57
6	0.1157	5.78
7	0.0925	4.63
8	0.2754	8.91

(c) The χ^2 goodness of fit test tests H_0: The number of revertants has a Poisson distribution versus H_1: The number of revertants does not have a Poisson distribution. The test statistic is

$$\chi^2 = \sum_{k=1}^{8} \frac{\left(f_k - \hat{f}_k\right)^2}{\hat{f}_k}$$
$$= \frac{(3-6.42)^2}{6.42} + \frac{(5-4.86)^2}{4.86} + \frac{(6-6.11)^2}{6.11} + \cdots + \frac{(5-8.91)^2}{8.91}$$
$$= 5.061.$$

There are $8 - 1 - 1 = 7$ degrees of freedom, so that the $p-$value $= 0.536$. Thus, there is not enough evidence to say that the data does contradict the claim that the number of revertants has a Poisson distribution.

(d) Two of the eight categories have expected frequencies, $\hat{f}_k$, less than five, i.e. 75% of the categories do have expected frequencies greater than five. Also, no expected frequency is zero. The data does satisfy these Rules of Thumb for large samples.

6.13. (a) Let π_1 denote the proportion of leaf and stem infested plants in the untreated population and π_2 denote the proportion of leaf and stem infested plants in the treated population. From Example 6.10,

$$\hat{\pi}_1 = 83/100 = 0.83 \qquad \hat{\pi}_2 = 33/100 = 0.33.$$

Then, $\hat{\theta} = \hat{\pi}_1 - \hat{\pi}_2 = 0.83 - 0.33 = 0.50$ and

$$\hat{\sigma}_{\hat{\theta}} = \sqrt{\frac{0.83(1-0.83)}{100} + \frac{0.33(1-0.33)}{100}} = 0.0602.$$

The 95% lower bound is

$$\hat{\theta} - z(0.05)\hat{\sigma}_{\hat{\theta}} = 0.50 - 1.645(0.0602) = 0.4010.$$

(b) With 95% confidence the difference in proportions of leaf and stem infested plants in the untreated and treated populations is at least 0.4010. In other words, the proportion of leaf and stem infested plants in the untreated population is at least 0.4010 more than the proportion of leaf and stem infested plants in the treated population.

Because the lower bound does not include zero, we can conclude at the 0.05 significance level that treated tobacco plants have less infection than untreated plants.

(c) The claim is equivalent to saying that $\pi_1 = 0.80$. The hypotheses test is

H$_0$: $\pi_1 = 0.80$

H$_1$: $\pi_1 \neq 0.80$

Test Statistic: $z_c = \dfrac{\hat{\pi}_1 - \pi_0}{\hat{\sigma}_{\hat{\pi}}} = \dfrac{0.83 - 0.80}{\sqrt{\dfrac{0.80(1-0.80)}{100}}} = 0.75$

From Table C.1 we have p−value $= 0.2266$

Conclusion: This p−value is large. There is not enough evidence to contradict the claim that the ratio is 80:20.

6.15. (a) Let π_1 denote the proportion of females with low self-esteem and π_2 denote the proportion of males with low self-esteem. From the table we have

$$\hat{\pi}_1 = \frac{42}{100} = 0.42 \qquad \hat{\pi}_2 = \frac{30}{105} = 0.28.$$

Let $\theta = \pi_1 - \pi_2$, then $\hat{\theta} = \hat{\pi}_1 - \hat{\pi}_2 = 0.42 - 0.28 = 0.14$ and

$$\hat{\sigma}_{\hat{\theta}} = \sqrt{\frac{\hat{\pi}_1(1-\hat{\pi}_1)}{n_1} + \frac{\hat{\pi}_2(1-\hat{\pi}_2)}{n_2}} = \sqrt{\frac{0.42(1-0.42)}{100} + \frac{0.28(1-0.28)}{105}}$$

$$= 0.066.$$

The hypothesis test is

H$_0$: $\theta \leq 0$

H$_1$: $\theta > 0$

Test Statistic: $z_c = \dfrac{0.14 - 0}{0.066} = 2.1212$ which gives p−value $= 0.0170$

Conclusion: Because the p−value is small, there is sufficient evidence that the proportion of females with low self-esteem is higher than the proportion of males with low self-esteem.

(b) The $\chi^2()$ test of the equality of distribution of these two categorical populations (self-esteem levels for males and females) is as follows. First, the following table of marginal values and estimated expected frequencies is calculated.

| Gender | Self-Esteem Level | | | |
	Low	Medium	High	f_{i+}
Male	$\hat{f}_{11} = 36.3878$	54.2927	13.8293	105
Female	35.1219	51.7073	13.1707	100
f_{+k}	72	106	27	205

where $\hat{f}_{11}$, the estimated expected frequency for males with low self-esteem, is calculated as

$$\hat{f}_{11} = \frac{f_{1+}f_{+1}}{f_{++}} = \frac{(105)(72)}{205} = 36.8780.$$

The other estimated expected frequencies are calculated similarly. Finally, the test is

H$_0$: $\pi_{1k} = \pi_{2k} \quad k = 1, 2, 3$

H$_1$: Not all $\pi_{1k} = \pi_{2k} \quad k = 1, 2, 3$

Test Statistic:

$$\chi^2(c) = \sum_{i=1}^{2} \sum_{k=1}^{3} \frac{f_{ik}^2}{\hat{f}_{ik}} - f_{++} = \frac{30^2}{36.868} + \frac{56^2}{54.2927} + \cdots + \frac{8^2}{13.1707} - 205 = 6.4805$$

Comparing this to a $\chi^2(2)$ distribution gives p−value between 0.025 and 0.05

Conclusion: A 0.05 level test would reject the null hypothesis, but a 0.025 level test would not. Being more precise would require the use of software rather than the tables.

6.17. The data can be arranged as in the following table.

	Germinate	Not Germinate	
Variety 1	6	4	10
Variety 2	3	7	10
	9	11	20

The hypotheses are H_0: $\pi_{11} = \pi_{21}$ versus H_1: $\pi_{11} > \pi_{21}$. The p-value can be calculated as $p = \sum_{y=6}^{10} f(y|9, 10, 20)$ where

$$f(y|9, 10, 20) = \frac{m_1!m_2!n_1!n_2!}{n!a!b!c!d!} = \frac{9!11!10!10!}{20!y!4!3!7!}.$$

It is much better to leave these kinds of calculations to the computer. The abbreviated output from PROC FREQ in SAS gives

```
Statistic                      DF    Value      Prob
------------------------------------------------------
Fisher's Exact Test (Left)                      0.965
                    (Right)                     0.185
                    (2-Tail)                    0.370
```

We are interested in the right-tailed (upper-tailed) p-value. Because this is 0.185 which is bigger than $\alpha = 0.05$, say, we do not reject H_0. The data does not support the conclusion at the 0.05 level that the germination rate for variety 1 is higher than that for variety 2.

6.19. Let π_1 denote the proportion of consumers who like Brand A ice cream and π_2 denote the proportion of consumers who like Brand B ice cream. Let $\theta = \pi_1 - \pi_2$. Because $\hat{\pi}_1$ and $\hat{\pi}_2$ are not independent (each consumer answered both questions) McNemar's Test is appropriate.

$$\hat{\theta} = \frac{B - C}{N} = \frac{92 - 13}{200} = 0.395$$

and

$$\hat{\sigma}_{\hat{\theta}} = \frac{\sqrt{B + C}}{N} = \frac{\sqrt{92 + 13}}{200} = 0.0512.$$

Then the 99% confidence interval for θ is

$$\hat{\theta} \pm z(0.01/2)\hat{\sigma}_{\hat{\theta}} = 0.395 \pm 2.55(0.0512) = (0.2644, 0.5256).$$

With 95% confidence the difference between the proportion of consumers who like Brand A and the proportion who like Brand B ice creams is between 0.2644 and 0.5256. Because this interval does not contain zero, we can conclude, with 95% confidence, that the proportion of consumers who like Brand A ice cream is statistically significantly different from the proportion of consumers who like Brand B.

6.21. The observations are in case-control pairs so the McNemar test can be used. Let π_1 be the proportion of estrogen users among cases and π_2 be the proportion of estrogen users among controls and $\theta = \pi_1 - \pi_2$. Then from the table

$$\hat{\theta} = \frac{B - C}{N} = \frac{65 - 9}{187} = 0.2995$$

50

and

$$\hat{\sigma}_{\hat{\theta}} = \frac{\sqrt{B+C}}{N} = \frac{\sqrt{65+9}}{187} = 0.0460.$$

The 99% lower bound is

$$\hat{\theta} - z(0.01)\hat{\sigma}_{\hat{\theta}} = 0.2995 - 2.33(0.0460) = 0.1923.$$

With 99% confidence the proportion of estrogen users among cases is at least 0.1923 larger than the proportion of estrogen users among controls.

Chapter 7

Designing Research Studies

7.1. (a) **Design 1:** Experimental Study, because the researcher controls the levels of OC user status and age.

Design 2: Experimental Study, because the researcher controls the levels of OC user status. Age is not controlled.

Design 3: Observational Study, because the research merely observes the values of OC user status and age without controlling the occurrence of either.

(b) For all three designs the response variable is Bacteriuria state which is a 1 if the bacteriuria is present and 0 if not. This is the response because it is thought to be affected by the other variables.

The explanatory variables for all three designs are the OC user status, which could be a 1 if the person uses oral contraceptives and a 0 if not, and age group. These are explanatory variables because they are thought to affect the likelihood of bacteriuria being present.

(c) The response variable is nominal: present (1) or absent (0). There is no ordering to the levels.

The OC user status is also nominal: OC user (1) or OC non-user (0). Again, there is no ordering to the levels.

The age group is ordinal because there is a natural ordering to the groups, 16-19, 20-29, 30-39, and 40-49. However, the levels of the variable can't be subtracted so age group is not interval.

(d) The factor is OC user status. The experimenters want to know if OC user status affects the likelihood of bacteriuria being present. This is the primary objective of the study.

Age is a confounding variable. It is thought to affect the response, but it is not of primary concern.

(e) The 200 females are the observational units because they each are the smallest units on which the bacteriuria status is measured.

7.3. (a) This is an experimental study because the researchers controlled the levels of daily calorie intake and the levels of daily exercise. The researcher also controlled gender by selecting 12 males and 12 females.

(b) The response is $Y = (Y_1, Y_2, Y_3)$ where Y_1, Y_2, and Y_3 are, respectively, the 4, 8, and 12 months weights of an experimental subject.

(c) **Gender** (Male, Female) qualitative
Calorie Intake (Low, High) qualitative
Daily Exercise (Low, High) qualitative

(d) Because they are thought to affect a subject's weight, the subjects age and/or bone structure are confounding variables. The subject's age is quantitative. The bone structure is qualitative with ordinal levels small, medium, and large.

(e) The treatments are the eight combinations of gender, calorie intake, and daily exercise. For example, one treatment is (Male, Low, Low).

(f) The experimental units are the subjects. The observational units are the subjects at the three time periods- 4, 8, and 12 months after the start of the treatment. The treatments are applied to the subjects and the responses are measured on the subjects at each time period.

7.5. (a) There are six treatments, (A, I), (A, O), (B, I), (B, O), (C, I), and (C, O). Thus, we need six subjects randomized to the six treatments in each block for a total of 24 subjects. Numbering the subjects from one to six, the experimental layout might look like

Block	(A, I)	(A, O)	(B, I)	(B, O)	(C, I)	(C, O)
1	1	3	6	2	4	5
2	4	6	5	2	3	1
3	2	5	3	4	1	6
4	3	2	5	1	4	6

(b) There are four replications for each treatment.

(c) One way to randomize the assignment of treatments to experimental units is to number the subjects in each block from 1 to 6. Place six numbered slips of paper in a hat and mix well. Select the slips from the hat and assign the subject selected first to (A, I), the subject selected second to (A, O), and so on. Repeat this for each block.

7.7. Let Y_{ij} denote the time taken for relief for the jth subject given the ith treatment and suppose that subjects were confounding. Then, for design 1, the model is

$$Y_{ij} = \mu_i + \rho_j + E_{ij}$$

where $j = 1, 2, \ldots, 20$ if $i = 1$ and $j = 11, 12, \ldots, 20$ if $i = 2$. The layout for design 1 would be

Group	Subject	Response	Expected Response
Treated	1	$Y_{1,1}$	$\mu_1 + \rho_1$
Treated	2	$Y_{1,2}$	$\mu_1 + \rho_2$
Treated	3	$Y_{1,3}$	$\mu_1 + \rho_3$
$\vdots$	$\vdots$	$\vdots$	$\vdots$
Treated	10	$Y_{1,10}$	$\mu_1 + \rho_{10}$
Untreated	11	$Y_{2,11}$	$\mu_2 + \rho_{11}$
Untreated	12	$Y_{2,12}$	$\mu_2 + \rho_{12}$
Untreated	13	$Y_{2,13}$	$\mu_2 + \rho_{13}$
$\vdots$	$\vdots$	$\vdots$	$\vdots$
Untreated	20	$Y_{2,20}$	$\mu_2 + \rho_{20}$

The expected response of comparisons between subjects in the two groups would be, for example,

$$(\mu_1 + \rho_1) - (\mu_2 + \rho_{11}) = (\mu_1 - \mu_2) + (\rho_1 - \rho_{11}).$$

Thus, the difference between the group means could not be separated from the effect of the subject, i.e. the confounding variable.

The model for design 2, on the other hand, would be

$$Y_{ij} = \mu_i + \rho_j + E_{ij} \qquad i = 1,2; j = 1,2,\ldots,10.$$

and the layout

Subject	Eye	Group	Response	*Expected Response*
1	1	Treated	$Y_{1,1}$	$\mu_1 + \rho_1$
1	2	Untreated	$Y_{2,1}$	$\mu_2 + \rho_1$
2	1	Treated	$Y_{1,2}$	$\mu_1 + \rho_2$
2	2	Untreated	$Y_{2,2}$	$\mu_2 + \rho_2$
$\vdots$	$\vdots$	$\vdots$	$\vdots$	$\vdots$
10	1	Treated	$Y_{1,10}$	$\mu_1 + \rho_{10}$
10	2	Untreated	$Y_{2,10}$	$\mu_2 + \rho_{10}$

Here the expected response of the comparisons would be, for example,

$$(\mu_1 + \rho_1) - (\mu_2 + \rho_1) = \mu_1 - \mu_2.$$

This design accounts for the confounding variable so that the comparisons are free of its effect. Thus, if the subject were a confounding variable, then design 2 would be preferred.

If the subject were not a confounding variable, then design 2 might still be preferred, because it requires fewer subjects. However, it might be more difficult to perform and losing a subject means losing two observations.

7.9. From Example 4.8 the pooled standard deviation is 3.34. Using the formula from Box 7.1, n is the smallest integer such that

$$n \geq a\left[t(\nu, b\alpha) + t(\nu, \beta)\right]^2 \left(\frac{\sigma}{\Delta}\right)^2$$

In this inequality we have $a = 2$ because we have two samples and $b = 1$ because it is an upper-tailed test, $\Delta = 5\text{ppm}$, $\alpha = 0.05$, and $\beta = 0.10$. Then n is the smallest integer such that

$$n \geq \left[t(\nu, 0.05) + t(\nu, 0.01)\right]^2 \left(\frac{3.24}{5}\right)^2$$

Through trial and error we arrive at the following table

n	ν	RHS
5	8	9.4634
10	18	8.3809
8	14	8.6115
9	16	8.4806

Thus, $n = 9$ is the smallest integer that meets the criterion and therefore taking nine water samples in both locations will ensure with 90% confidence that if the actual difference is more than 5ppm, then an 0.05 level independent t-test will conclude that the mean pollution level downstream is higher than the mean pollution level upstream.

7.11. From Box 7.4, the sample size is the smallest integer such that

$$n \geq b\left[t(\nu, \alpha/2)\right]^2 \left(\frac{\sigma}{B}\right)^2$$

55

In this inequality $b = 2$ because we are looking at the difference between the true mean chemical levels in the downstream and upstream waters, $\alpha = 0.05$, and $B = 2$ppm. Then, n is the smallest integer such that

$$n \geq 2 \, [t(\nu, 0.025)]^2 \left(\frac{3.24}{2}\right)^2$$

Table C.2 in the appendix only gives the values for certain degrees of freedom. Using a statistical package like S-plus allows us to find the exact sample size that satisfies the above inequality.

n	ν	$t(\nu, \alpha/2)$	RHS
19	36	2.0281	21.5892
20	38	2.0244	21.5105
21	40	2.0211	21.4400
22	42	2.0181	21.3765
23	44	2.0154	21.3191

Therefore, $n = 22$ is the smallest value that is greater than or equal to the right hand side (RHS).

7.13. We are told that 180 seeds germinated out of 210 cultivated in medium I ($\hat{p} = 0.857$) and that we want a margin of error of less than 5%. The value of $\hat{\sigma} = 0.50$ will be used to give a conservative estimate. Because this is one-sample estimation we have $b = 1$. Then, assume $\alpha = 0.10$ and the sample size is

$$n = b \, [z(0.05)]^2 \left(\frac{\sigma}{B}\right)^2 = 1 \, [1.645]^2 \left(\frac{0.5}{.05}\right)^2 = 270.6.$$

We round this up and use 271 seeds.

7.15. From Exercise **4.4** the estimated population standard deviation is 0.3517. From the text we have $\alpha = 0.01$ and $B = 0.5$. Then, n is the largest integer such that

$$n \geq b \, [t(\nu, 0.01/2)]^2 \left(\frac{0.3517}{0.5}\right)^2$$

where $b = 1$ because we are estimating a single population mean. Trying a few numbers gives the following table

n	ν	RHS
10	9	5.2257
7	6	6.8009
6	5	8.0444

Thus, $n = 7$ is the sample size that gives the desired conditions.

7.17. The goal is to use the hypothesis testing method to determine a sample size that meets the criterion stated in the Exercise. The criterion for hypothesis testing is usually stated in terms of the minimum distance that we would want to detect. In this exercise this minimum distance is expressed as a minimum probability. That is, "If there is a probability of at least 65% that the infection rate of a person selected at random from the study population exceeds the rate of a person randomly selected from a population with mean 1."

Let Y_1 be the infection rate of a person selected at random from the study population and Y_2 be the infection rate of a person selected at random from a population with mean 1.

Using Box 7.3,

$$Pr\{Y_1 \geq Y_2\} = 0.65 = A\left(-\frac{\Delta}{\sigma\sqrt{2}}\right)$$

From Table C.1

$$-\frac{\Delta}{\sigma\sqrt{2}} \approx -0.38$$

$$\Rightarrow \frac{\Delta}{\sigma} \approx 0.5374$$

Finally, n is the smallest integer such that

$$n \geq a\left[t(\nu, b\alpha) + t(\nu, \beta)\right]^2 \left(\frac{\sigma}{\Delta}\right)^2$$

where $a = 1$ and $b = 1$, because this is an upper-tailed test, and $\alpha = 0.05$ and $\beta = 0.20$. Then, we have

$$n \geq [t(\nu, 0.05) + t(\nu, 0.20)]^2 \, 1.8608.$$

Trying a few values of n gives the following table

n	ν	RHS
20	19	23.2299
25	24	22.8316
23	22	22.9670
22	21	23.0455

Therefore, $n = 23$ is the desired sample size.

Chapter 8

Single Factor Studies: One-Way ANOVA

8.1. Design 1 is a completely randomized design. The treatments are assigned randomly and independently to subjects. Design 2, on the other hand, does not have an independent assignment. There are two treatments in this design (treatment, no treatment). [Note that the word treatment is used twice with different meanings. This is a common, and unfortunate, occurrence. It leads to silly statements like "The subjects in the second treatment received no treatment." It is hoped that the meaning can be understood from the context.] There are 10 replications per treatment.

8.3. (a)

$$
\begin{aligned}
Y_{1+} &= 1.59 + 1.73 + 3.64 + 1.97 = 8.93 \\
Y_{2+} &= 3.36 + 4.01 + 3.49 + 2.89 = 13.75 \\
Y_{3+} &= 3.92 + 4.82 + 3.87 + 5.39 = 18.00
\end{aligned}
$$

(b)

$$
\begin{aligned}
Y_{++} &= Y_{1+} + Y_{2+} + Y_{3+} = 8.93 + 13.75 + 18.00 = 40.68 \\
\overline{Y}_{++} &= \frac{Y_{++}}{12} = \frac{40.68}{12} = 3.39
\end{aligned}
$$

(c)

$$
\begin{aligned}
\overline{Y}_{1+} &= \frac{Y_{1+}}{4} = \frac{8.93}{4} = 2.2325 \\
\overline{Y}_{2+} &= \frac{Y_{2+}}{4} = \frac{13.75}{4} = 3.4375 \\
\overline{Y}_{3+} &= \frac{Y_{3+}}{4} = \frac{18.00}{4} = 4.50
\end{aligned}
$$

(d)

$$
\begin{aligned}
S_1^2 &= \frac{(1.59 - 2.2325)^2 + (1.73 - 2.2325)^2 + (3.64 - 2.2325)^2 + (1.97 - 2.2325)^2}{4 - 1} \\
&= 0.9051 \\
S_2^2 &= \frac{(3.36 - 3.4375)^2 + \ldots + (2.89 - 3.4375)^2}{4 - 1} = 0.2121 \\
S_3^2 &= \frac{(3.92 - 4.5)^2 + \ldots + (5.39 - 4.5)^2}{4 - 1} = 0.5426
\end{aligned}
$$

8.5. (a) Plots for the observations with sample means(Fig(a)), and sample means with the overall mean(Fig(b)) follow.

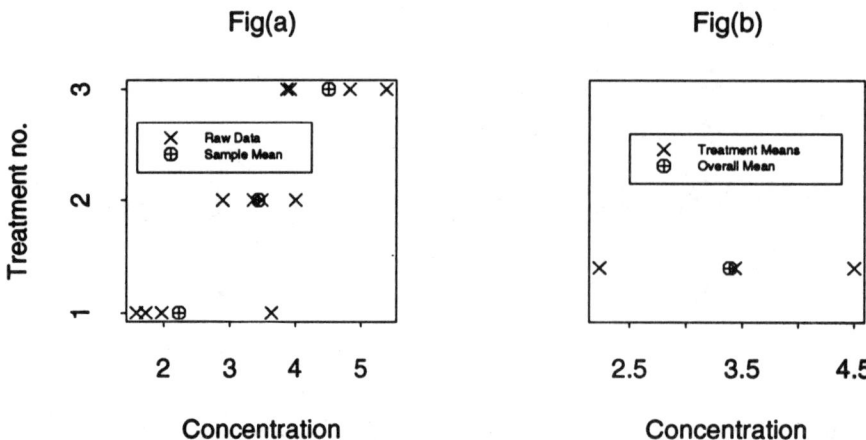

(b) For Treatment 1 there seems to be a bit more variability than for the other two Treatments. This seems to be due to a single observation that could well be an outlier. Because of this it seems that the amount of variability for each of the samples is similar. Thus, it is reasonable to assume that the variances are the same for the three populations.

(c) A reasonable estimate of σ^2 would be a weighted average of the S_i^2. In this case, all sample sizes are equal, so a weighted average is just the sample average.

$$\hat{\sigma}^2 = \frac{S_1^2 + S_2^2 + S_3^2}{3} = \frac{.9051 + .2121 + .5426}{3} = .5533$$

<u>NOTE</u>: This estimate corresponds to $MS[E]$.

(d) In looking at the sample plots it seems that the sample means are very far apart. This would imply that the null hypothesis $H_0 : \mu_1 = \mu_2 = \mu_3$ would be rejected.

(e) If H_0 is rejected, at least two of the population means are different. In other words, at least two of the three levels of glucose cause different average amounts of insulin to be released by the patients.

(f) For this data set the assumptions correspond to the following:

    i. The samples from the three treatments of glucose levels are independent.

    ii. The populations of amount of insulin released for the patients have normal distributions.

    iii. The variances of all three populations are identical.

(g) $MS[E]$ was actually calculated in Part Part **c** in Exercise **8.5** of this problem. $MS[T]$ can be found as follows.

$$MS[T] = \frac{n}{t-1}[(\overline{Y}_{1+} - \overline{Y}_{++})^2 + (\overline{Y}_{2+} - \overline{Y}_{++})^2 + (\overline{Y}_{3+} - \overline{Y}_{++})^2]$$

$$= \frac{4}{3-1}[(2.2325 - 3.39)^2 + (3.4375 - 3.39)^2 + (4.5 - 3.39)^2] = 5.1483.$$

F_c can then be calculated as

$$F_c = \frac{MS[T]}{MS[E]} = \frac{5.1483}{.5533} = 9.30,$$

which is compared to a $F(2,9)$. The associated p-value is $< .01$, hence, the null hypothesis is rejected. The statement in Part Part **e** in Exercise **8.5** of this problem is then concluded.

8.7. Preliminary calculations give us

$$\overline{Y}_{0+} = 1.5833, \quad \overline{Y}_{400+} = 1.6917, \quad \overline{Y}_{800+} = 1.8667, \quad \overline{Y}_{++} = 1.7139,$$

$$S_0^2 = .0013, \quad S_{400}^2 = .0143, \quad \text{and} \quad S_{800}^2 = .0407.$$

(a) Plots for the observations with sample means(Fig(a)), and sample means with the overall mean(Fig(b)) follow.

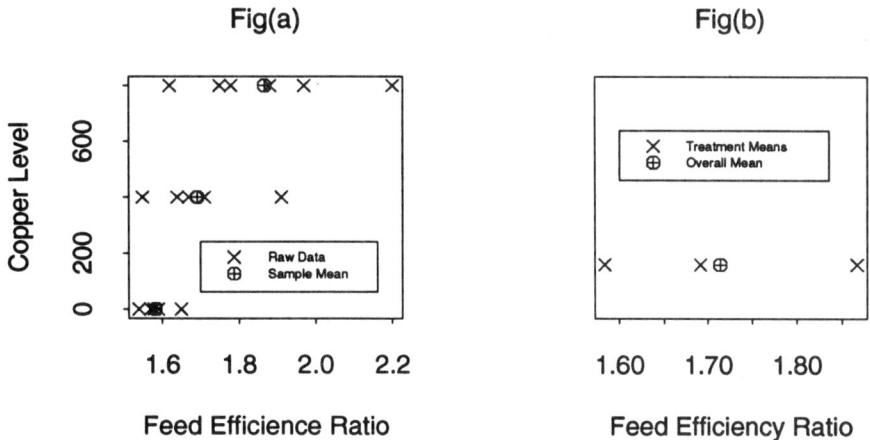

(b) The amount of variability for each of the samples seems different. In fact, the amount of variability seems to increase with the Copper Level. The higher the Copper Level the more variability in the sample. Because of this, it does not seem reasonable that the variances are the same for the three populations of Feed Efficiency Ratios.

(c) A reasonable estimate of σ^2 would be a weighted average of the S_i^2. In this case all sample sizes are equal, so a weighted average is just the sample average.

$$\hat{\sigma}^2 = \frac{S_1^2 + S_2^2 + S_3^2}{3} = \frac{.0013 + .0143 + .0407}{3} = .0188$$

<u>NOTE</u>: This estimate corresponds to $MS[E]$, but in this case the variances may not be equal.

(d) In looking at the sample plots it seems that the sample means are very far apart. This would imply that the null hypothesis $H_0 : \mu_1 = \mu_2 = \mu_3$ would be rejected. There still remains a problem in that the variances do not seem to be equal.

(e) If H_0 is rejected, at least two of the population means are different. In other words, at least two of the three levels of copper cause different mean feed efficiency ratios at the end of three weeks.

(f) For this data set the assumptions correspond to the following:

    i. The samples from the three treatments of Copper Levels are independent.

    ii. The populations of feed efficiency ratios for the three treatments have normal distributions.

    iii. The amount of variability in the populations corresponding to the three Copper Levels are identical.

61

(g) $MS[E]$ was actually calculated in Part Part **c** in Exercise **8.7** of this problem. $MS[T]$ can be found as follows.

$$
\begin{aligned}
MS[T] &= \frac{n}{t-1}[(\overline{Y}_{0+} - \overline{Y}_{++})^2 + (\overline{Y}_{400+} - \overline{Y}_{++})^2 + (\overline{Y}_{800+} - \overline{Y}_{++})^2] \\
&= \frac{6}{3-1}[(1.5833 - 1.7139)^2 + (1.6917 - 1.7139)^2 + (1.8667 - 1.7139)^2] \\
&= 0.1227
\end{aligned}
$$

F_c can then be calculated as

$$
F_c = \frac{MS[T]}{MS[E]} = \frac{0.1227}{0.0188} = 6.5266,
$$

which is compared to a $F(2, 15)$. The associated p−value is $< .01$, hence, the null hypothesis is rejected. The statement in Part Part **e** in Exercise **8.7** of this problem is then concluded.

8.9. Using the calculations done in Exercise **8.3**, the following can be found.

$$
\begin{aligned}
CM &= \frac{40.68^2}{12} = 137.9052, \\
SS[TOT] &= 1.59^2 + ... + 5.39^2 - CM \\
&= 153.1812 - 137.9052 = 15.2760, \\
SS[T] &= \frac{8.93^2}{4} + \frac{13.75^2}{4} + \frac{18.00^2}{4} - CM \\
&= 148.2019 - 137.9052 = 10.2967 \\
SS[E] &= 15.2760 - 10.2967 = 4.9793.
\end{aligned}
$$

The ANOVA table has the following form.

Source	df	SS	MS	F_c
Treatment	2	10.2967	5.1483	9.31
Error	9	4.9793	0.5533	
Total	11	15.2760		

F_c is then compared to an $F(2, 9)$ and has associated p−value .0064. Since this value is quite small, it is reasonable to conclude that at least two of the three levels of glucose have different effects on the insulin release in patients.

8.11. (a) Since treatments were not actually applied to the patients, the study must be an observational study. The individuals in the study were not randomly assigned to the serum phosphate level. They were put into the groups because of their level and their mean responses were compared.

There is a single factor with three levels in this study. The three levels are

- Patients with sickle cell disease and high phosphate level.
- Patients with sickle cell disease and normal phosphate level.
- Normal patients.

(b) Preliminary calculations give the following:

$$
Y_{1+} = 23.21 \quad Y_{2+} = 14.30 \quad\quad Y_{3+} = 23.15 \quad Y_{++} = 60.66
$$

$$
CM = \frac{60.66^2}{16} = 229.9772,
$$

62

$$SS[TOT] = 5.36^2 + \ldots + 3.38^2 - CM$$
$$= 238.3198 - 229.9772 = 8.3426,$$
$$SS[T] = \frac{23.21^2}{5} + \frac{14.30^2}{4} + \frac{23.15^2}{7} - CM$$
$$= 235.4237 - 229.9772 = 5.4465,$$
$$SS[E] = 8.3426 - 5.4465 = 2.8961.$$

The ANOVA table has the following form.

Source	df	SS	MS	F_c
Treatment	2	5.4465	2.7232	12.22
Error	13	2.8961	.2228	
Total	15	8.3426		

F_c is then compared to an $F(2, 13)$ and has associated $p-$value.0010. Since the associated $p-$value is quite small, it is reasonable to conclude that at least two of the patient groups have different TRCP levels.

8.13. (a) Defining the 12 treatment means as $\mu_1, \ldots, \mu_{12}$, the null and alternative hypotheses can be written symbolically as

$$H_0 \ : \ \mu_1 = \mu_2 = \ldots = \mu_{12},$$
$$H_1 \ : \ \mu_i \neq \mu_j \text{ for at least one pair } i \neq j.$$

H_0 is the hypothesis that all twelve treatments(combinations of P-Release, Inoculation and Foliar P) have the same effect on leaf area ratios.

H_1 is the hypothesis that at least two of the twelve treatments have different effects on leaf area ratios.

(b) The data for the 12 distinct treatments are independent random samples from the corresponding populations.

The underlying population for each of the 12 treatments has a normal distribution.

The variances of the underlying populations are identical.

(c) The data are given as sample means of size three. From this the Y_{i+}'s are found by multiplying the means by three. The only difficulty is finding all of the SS's. This can be most easily accomplished by estimating $MS[E]$ and $SS[T]$. From these two, the rest of the table can be constructed. Since the treatments are equally replicated, $MS[E]$ is just the average of the sample variances.

Hence, preliminary calculations give the following:

$$MS[E] = \frac{\sum S_i^2}{12} = \frac{1.04^2 + 0.69^2 + \ldots + 1.04^2}{12} = .5711,$$
$$SS[E] = 24(MS[E]) = 13.7064$$
$$Y_{++} = 492.9,$$
$$CM = \frac{492.9^2}{36} = 6748.6225,$$
$$SS[T] = \frac{(3(11.6))^2}{3} + \ldots + \frac{(3(15.9))^2}{3} - CM$$
$$= \frac{34.8^2}{3} + \ldots + \frac{47.7^2}{3} - CM$$
$$= 6833.01 - 6748.6225 = 84.3875,$$
$$SS[TOT] = SS[E] + SS[T] = 13.7064 + 84.3875 = 98.0939.$$

The ANOVA table has the following form.

Source	df	SS	MS	F_c
Treatment	11	84.3875	7.6716	13.4330
Error	24	13.7064	.5711	
Total	35	98.0939		

F_c is then compared to an $F(11, 24)$ and has associated p-value$<.0001$. Since the associated p-value is quite small, it is reasonable to conclude that at least two of the twelve treatments have different effects on leaf area ratios.

8.15. (a) The necessary ANOVA assumptions are

    i. The feed efficiency ratios are independent random samples from the three populations.

    ii. The populations of feed efficiency ratios have normal distributions.

    iii. The populations of feed efficiency ratios have equal variances.

Using the calculations done in (**8.7**), the following can be found.

$$CM = \frac{(18\overline{Y}_{++})^2}{18} = \frac{(18(1.7139))^2}{18} = 52.8742,$$

$$SS[TOT] = 1.57^2 + \ldots + 2.20^2 - CM$$
$$= 53.4011 - 52.8742 = 0.5269,$$

$$SS[T] = \frac{(6(1.5833))^2}{6} + \frac{(6(1.6917))^2}{6} + \frac{(6(1.8667))^2}{6} - CM$$
$$= 53.1195 - 52.8742 = 0.2453,$$

$$SS[E] = 0.5269 - 0.2453 = 0.2816.$$

The ANOVA table has the following form

Source	df	SS	MS	F_c
Treatment	2	0.2453	0.1228	6.5412
Error	15	0.2816	0.0188	
Total	17	0.5269		

F_c is then compared to an F^2_{15} and has associated p-value < 0.01. We reject H_0.

(b) The research hypothesis is that at least two amounts of copper in the diet have different true mean feed efficiency ratios. Because we rejected the null hypothesis in favor of the research hypothesis, we conclude that at least two of the copper levels have different effects on feed efficiency ratio.

8.17. The data are given as means of samples, each of different size. From this the Y_{i+}'s are gotten by multiplying the sample mean by the sample size. The only difficulty is finding all of the SS's. This can be most easily accomplished by estimating $MS[E]$ and $SS[T]$. From these two, the rest of the ANOVA table can be constructed. Since the treatments are not equally replicated, $MS[E]$ is estimated as the weighted sum of the sample variances found in the text. Preliminary calculations give the following:

$$MS[E] = \frac{\sum(n_i - 1)S_i^2}{\sum n_i - t} = \frac{10(45)^2 + 8(19)^2 + \ldots + 20(24)^2}{103 - 7} = 1706,$$

$$SS[E] = 96(MS[E]) = 163776,$$

$Y_{1+} = 2475 \quad Y_{2+} = 1899\ldots \quad Y_{7+} = 3402 \quad Y_{++} = 20851$

$$CM = \frac{20851^2}{103} = 4221011.66,$$

$$SS[T] = \frac{2475^2}{11} + \frac{1899^2}{9} + \frac{3402^2}{21} - CM$$
$$= 4281787 - 4221011.66 = 60775.34,$$
$$SS[TOT] = SS[E] + SS[T] = 163776 + 60775.34 = 224551.34.$$

The ANOVA table has the following form.

Source	df	SS	MS	F_c
Treatment	6	60775.34	10129.22	5.93
Error	96	163776.00	1706.00	
Total	102	224551.34		

F_c is then compared to an $F(6, 96)$ and has associated p−value<.001. Since the associated p−value is quite small, it is reasonable to conclude that at least two of the groups have different levels of cholesterol.

8.19. Design 1 in Exercise 7.8 is the CRD. For this design the model can be written as

$$Y_{ij} = \mu + \tau_i + E_{ij} \qquad i = 1, 2, 3; j = 1, 2, \ldots, 8$$

where

Y_{ij} = the measured height of the jth plant in the ith group (time period),

μ = overall mean plant height,

τ_i = the effect of the ith time period; $\tau_1 + \tau_2 + \tau_3 = 0$,

E_{ij} = the random error; the E_{ij} are independent $N(0, \sigma^2)$.

8.21. A one-way ANOVA model for the data from this study can be specified as follows:

$$Y_{ij} = \mu + \tau_i + E_{ij} \quad i = 1, \ldots, 12; \; j = 1, 2, 3;$$

where,

Y_{ij} = measured leaf area ratio of the jth replicate for the ith treatment level.

μ = overall mean. That is, μ is the expected average leaf area ratio of the 36 leaves tested.

τ_i = the effect of the ith treatment. That is, τ_1 is the amount by which μ_1, the expected leaf area ratio of a leaf treated with low P-Release, no Foliar P and not inoculated deviates from μ. The τ_i's satisfy the condition: $\tau_1 + \ldots + \tau_{12} = 0$.

E_{ij} = the error of the jth replicate, for the ith treatment. That is, E_{ij} is the difference between the measured and expected leaf area ratio of the jth replicate in the ith treatment group. The E_{ij} are also assumed to be $N(0, \sigma^2)$.

8.23. The model for the feed efficiency data could be written as

$$Y_{ij} = \mu + \tau_i + E_{ij} \qquad i = 1, 2, 3; j = 1, 2, \ldots, 6$$

where

Y_{ij} = the feed efficiency ratio measured on the jth chick with the ith copper level.

μ = the overall mean. That is, μ is the expected feed efficiency for all chicks with these three copper levels.

τ_i = the effect of the ith copper level. That is, τ_i is the deviation of the expected feed efficiency for the ith copper level, μ_i, from overall mean, μ. The τ_i's satisfy $\tau_1 + \tau_2 + \tau_3 = 0$.

E_{ij} = the error in the jth response for the ith copper level. That is, E_{ij} is the difference between the measured and the expected feed efficiency ratio of the jth chick from the ith copper level.

8.25. Using the one-way ANOVA model in the box, first consider the sum

$$\sum_{j=1}^{n} Y_{ij} = \sum_{j=1}^{n} \mu + \sum_{j=1}^{n} \tau_i + \sum_{j=1}^{n} E_{ij} = n\mu + n\tau_i + E_{i+}$$

where $E_{i+} = E_{i1} + E_{i2} + \cdots + E_{in}$. Then the mean response for the ith treatment is

$$\frac{1}{n} \sum_{j=1}^{n} Y_{ij} = \frac{1}{n} \left(n\mu + n\tau_i + E_{i+} \right) = \mu + \tau_i + \overline{E}_{i+}$$

8.27. The residual vs. fitted and Normal probability plots for the insulin release data have the following form.

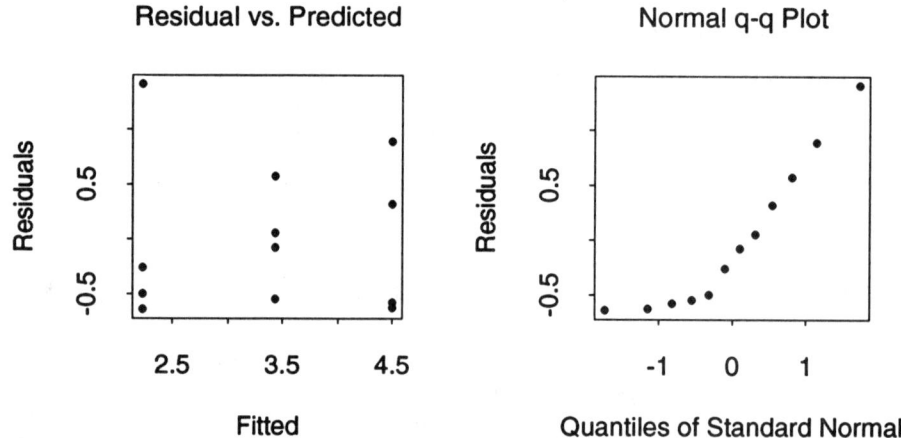

First, recognize that conclusions from residual plots based on a small amount of data should be interpreted with caution. With this in mind these plots indicate that the assumption of normality may not hold. The Normal probability plot indicates a severe departure from normality. The curvature of the Normal probability plot implies that the data have heavier tails than what is expected for the normal distribution. The assumption of equal variances is also suspect, since there seems to be more variability for the Low Concentration level than for the other two. However, this may be due to the one observation associated with Low Concentration of 3.64. This data point may be an outlier.

8.29. The residual vs. fitted and normal q-q plots for the TRCP data have the following form.

Residual vs. Predicted   ## Normal q-q Plot

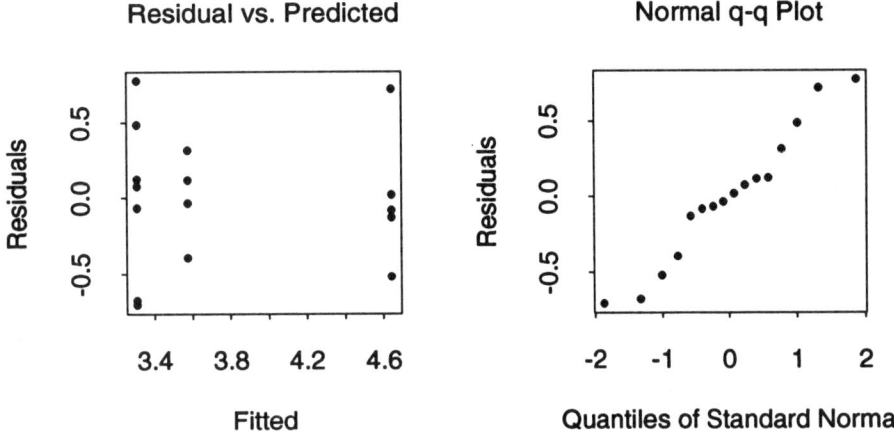

The ANOVA assumptions seem to be met here. One problem could be with equality of the variances. The subject group of Normal Serum Phosphate seems to have less variability than the other two groups. This may be due to fewer observations for the Normal Serum Phosphate Patients. The other problem is that the sample size is quite small and normality is difficult to check in this setting.

8.31. The residual vs. fitted and normal q-q plots for the plant height data have the following form.

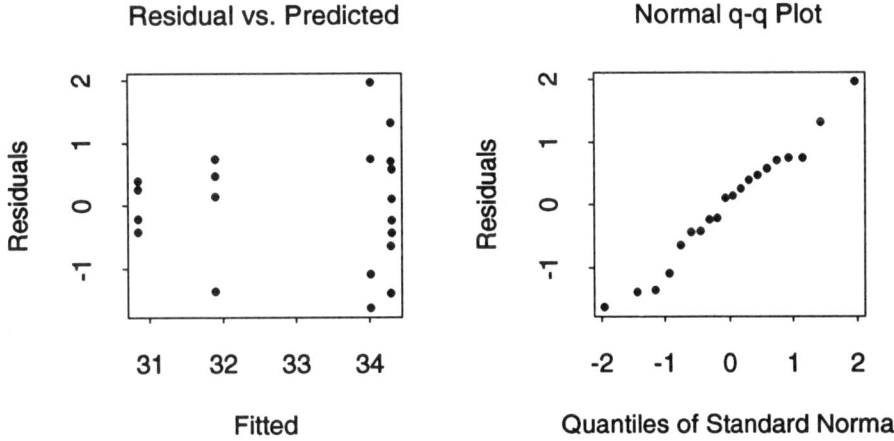

Based on the residual vs. fitted plot, there does seem to be evidence that a log-transformation may be useful. This is because the variability seems to be increasing with the mean, however this could be due to fewer observations associated with the smaller fitted values.

After taking the log transformation and refitting the data we get the following residual plots.

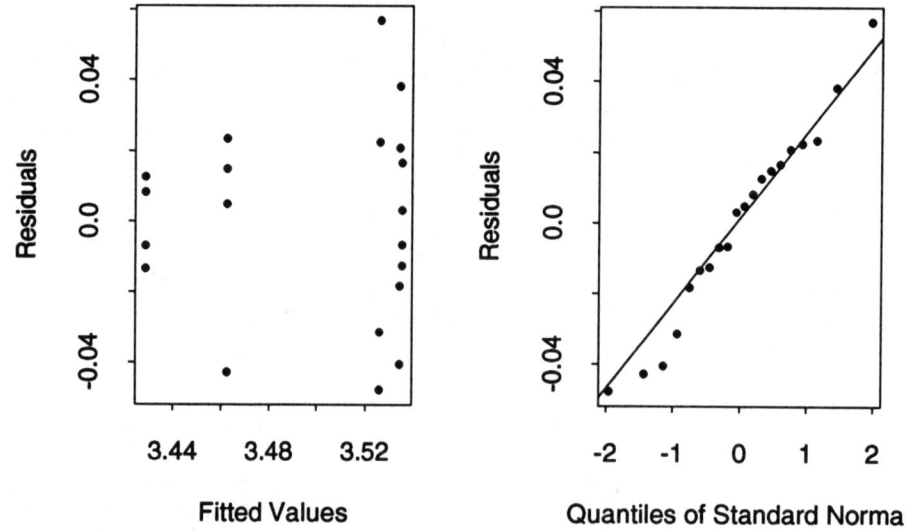

Based on these plots, the log transformation did not make the variances more consistent. It may be that another transformation may help more.

8.33. (a) A plot of the standard deviation versus the mean is

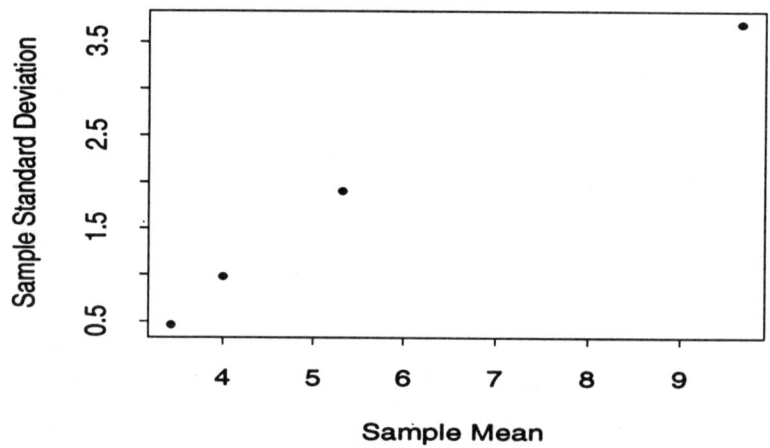

(b) After log-transforming the data, the residual plots are as shown below.

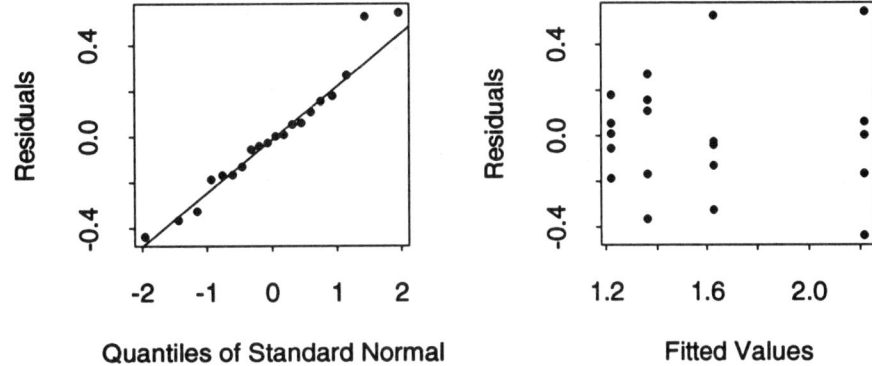

Both problems seem to have been alleviated somewhat. The points in the normal q-q plot fall closer to a straight line and the variability of the points stays more constant over the range of fitted values. However, there is still less variability for smaller fitted values than for larger fitted values.

8.35. (a) A plot of the sample means versus the sample standard deviations is given below.

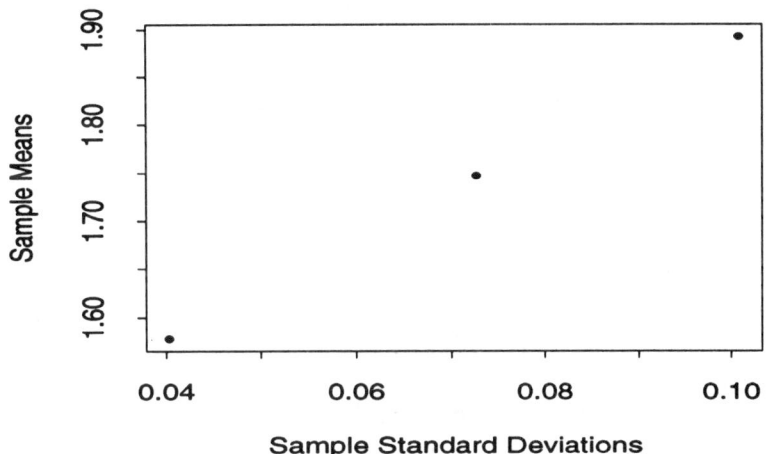

The relationship does seem to be linear. Therefore, a log-transformation might improve the validity of the ANOVA assumption; specifically, the equal variances assumption.

(b) The normal q-q plot of the residuals of the log-transformed data and the plot of the residuals versus the fitted values for the log-transformed data are shown below.

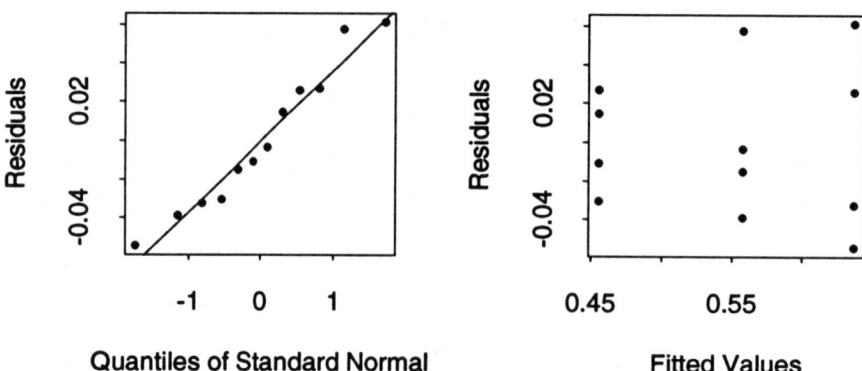

The behavior of the residuals has improved over the behavior of the residuals of the untransformed data.

8.37. (a) The test statistic for the Kruskal-Wallis test is

$$K_1 = \frac{12}{N(N+1)} \sum_{i=1}^{t} \frac{R_{i+}^2}{n_i} - 3(N+1)$$

$$= \frac{12}{12(13)} \left[\frac{13^2}{4} + \frac{25^2}{4} + \frac{40^2}{4} \right] - 3(13)$$

$$= 7.0385.$$

Note that $C = 1$ since there are no ties (i.e. all tied groups are of size 1).

Then the test is

H₀: The population distributions are all equal

H₁: The population distributions are not all equal with respect to location

Test Statistic: $K = \dfrac{K_1}{C} = 7.0385$

Rejection Region: From Table C.9 the critical value for an $\alpha = 0.056$ test is 5.6923. Hence, reject H₀ if $K > 5.6923$.

Conclusion: At the 0.05 level, the data does show that the true median insulin releases are not the same at the three concentrations.

(b) A model for the insulin data could be written as

$$Y_{ij} = \mu + \tau_i + E_{ij}, \qquad j = 1, 2, 3, 4; i = 1, 2, 3$$

where

Y_{ij} is the insulin release of the jth pancreatic tissue in the ith concentration,

μ is the overall median insulin released,

τ_i is the effect of the ith concentration, i.e. the difference between μ and the true median insulin release for the ith concentration, and

E_{ij} is the random error, i.e. the difference between the jth insulin release and the expected median insulin release for the ith concentration. The E_{ij} have a common continuous distribution.

(c) Since we are assuming that the population distributions are equal except for location, we can say that there is a difference in the true mean MCI values for the three concentrations.

(d) These results are the same as those obtained in the ANOVA test in Exercise **8.9**. The consistency of the tests lend strength to the validity of their results. If the assumptions for the ANOVA test were not met, then it is possible that the Kruskal-Wallis would give different results.

8.39. The Kruskal-Wallis test statistic begins with

$$K_1 = \frac{12}{N(N+1)} \sum_{i=1}^{t} \frac{R_{i+}^2}{n_i} - 3(N+1)$$

$$= \frac{12}{16(17)} \left[\frac{70^2}{5} + \frac{28^2}{4} + \frac{38^2}{7} \right] - 3(17)$$

$$= 9.9832.$$

and, again, $C = 1$. The test is conducted as follows

H_0: The three population distributions are equal

H_1: The three population distributions are not equal

Test Statistic: $K = \dfrac{K_1}{C} = 9.9832$

Rejection Region: Reject H_0 if $K > \chi^2(2, 0.01) = 9.21034$

Conclusion: At the $\alpha = 0.01$ level there is sufficient evidence to say that the population TRCP distributions vary depending upon serum phosphate level.

8.41. (a) Since we are working with ranks (ratings), an appropriate test would be the Kruskal-Wallis. To calculate this we need

$$K_1 = \frac{12}{N(N+1)} \sum_{i=1}^{t} \frac{R_{i+}^2}{n_i} - 3(N+1)$$

$$= \frac{12}{9(10)} \left[\frac{9^2}{3} + \frac{24^2}{3} + \frac{12^2}{3} \right] - 3(10)$$

$$= 5.6.$$

Since there are no ties, $C = 1$. The hypothesis test is

H_0: The three population distributions are equal

H_1: The three population distributions are not equal

Test Statistic: $K = 5.6$

Comparing this to the critical values in Table C.3 we find that the $p-$value is 0.05.

Conclusion: With a $p-$value of 0.05, we would reject at the 0.05 significance level and conclude that the median rating is different for the three colors.

(b) The null and research hypotheses are H_0: The three population distributions are equal and H_1: The three population distributions are not equal. We are assuming that the ranks are the result of ranking three independent samples of palatability scores.

8.43. The ranks of the survival times and the sum by treatment are

	1	2	3	4
	1.5	6.0	1.5	6.0
	6.0	21.5	6.0	21.5
	6.0	21.5	6.0	21.5
	21.5	21.5	6.0	21.5
	21.5	21.5	21.5	21.5
	21.5	21.5	21.5	21.5
	21.5	21.5	21.5	42.0
	21.5	21.5	21.5	42.0
	34.0	42.0	21.5	42.0
	42.0	42.0	21.5	51.5
	42.0	42.0	21.5	61.0
	42.0	51.5	42.0	61.0
	42.0	54.0	42.0	61.0
	42.0	55.5	42.0	64.0
	42.0	55.5	51.5	61.0
	51.5	57.0	61.0	58.0
R_{i+}	458.5	556.0	408.5	657.0

Then,

$$K_1 = \frac{12}{N(N+1)} \sum_{i=1}^{t} \frac{R_{i+}^2}{n_i} - 3(N+1)$$

$$= \frac{12}{64(65)} \left(\frac{458.5^2}{16} + \frac{556.0^2}{16} + \frac{408.5^2}{16} + \frac{657.0^2}{16} \right) - 3(65)$$

$$= 7.0142.$$

There are twelve groups of tied rangs with the tie sizes as

Group	66	72	95	98	116	138	143	163	192	274	360	368
Size	2	7	24	1	15	4	1	2	1	1	5	1

From this

$$C = 1 - \frac{\sum_{j=1}^{g} t_j(t_j^2 - 1)}{N^3 - N}$$

$$= 1 - 0.0675$$

$$= 0.9325.$$

so that the test statistic for comparing the four treatment population's survival distributions becomes

$$K = \frac{K_1}{C} = \frac{7.0142}{0.9325} = 7.5219.$$

We compare this to $\chi^2(3, .01) = 11.3449$. Hence, at the 0.01 significance level, there is not sufficient evidence to indicate that the four survival distributions are different. Note, however, that for someone who had chosen a 0.10 significance level, the results would be significant.

The large number of ties is evidence that the data comes from a non-normal distribution. Therefore, the Kruskal-Wallis Rank Test with the adjustment for ties would be more appropriate than the ANOVA F-test.

8.45. (a) We can write the jth response for treatment i as

$$Y_{ij} = \begin{cases} 1 & \text{if the } j^{th} \text{ site in the } i^{th} \text{ county was predominantly pine,} \\ 2 & \text{if the } j^{th} \text{ site in the } i^{th} \text{ county was predominantly hardwood,} \\ 3 & \text{if the } j^{th} \text{ site in the } i^{th} \text{ county was predominantly swamp.} \end{cases}$$

(b) There are four counties (populations) so $R = 4$, and three terrain types (categories) so $C = 3$. f_{12} is the frequency in county 1 and terrain type 2. From the table $f_{12} = 16$. Similarly, $f_{33} = 28$. f_{2+} is the total frequency in county 2 which is 50. f_{+3} is the total frequency in terrain type 3 which is 83. Finally, f_{++} is the total frequency of all counties and all terrain types which is 200.

(c) The null hypothesis means that the probability that a randomly selected site is a particular terrain type is the same for all four counties.

(d) The usual alternative hypothesis is

$$H_1 : \pi_{ik} \neq \pi_{jk} \text{ for at least one pair } i \neq j$$

This can be interpreted as the probability that a randomly selected site is a particular terrain type is different for at least one terrain type in at least two counties.

(e) The expected frequency for population 1 in category 1 is

$$\hat{f}_{11} = \frac{(50)(73)}{200} = 18.25.$$

The table of expected frequencies is

County #	Predominantly Pine	Predominantly Hardwood	Predominantly Swamp
1	18.25	11	20.75
2	18.25	11	20.75
3	18.25	11	20.75
4	18.25	11	20.75

Then

$$\chi_c^2 = \sum \frac{f_{ik}^2}{\hat{f}_{ik}} - f_{++}$$
$$= \frac{9^2}{18.25} + \frac{25^2}{18.25} + \frac{14^2}{18.25} + \cdots + \frac{14^2}{20.75} - 200$$
$$= 20.8125$$

We reject H_0 if $\chi_c^2 \geq \chi^2(6, .05) = 12.5916$. Since it is, we can conclude that the data shows that the probability that a randomly selected site is a particular terrain type is different for at least one terrain type in at least two counties.

8.47. The table is completed as in the text, with the expected frequencies in parenthesis. The expected frequencies are found by the following equation

$$\hat{f}_{ik} = \frac{f_{i+}f_{+k}}{f_{++}}.$$

For example

$$\hat{f}_{11} = \frac{f_{1+}f_{+1}}{f_{++}} = \frac{50(80)}{150} = 26.6667.$$

Treatment	Survivors after 8 days	Deaths in first 8 days	Total
I	20(26.6667)	30(23.3333)	50
II	25(26.6667)	25(23.3333)	50
III	35(26.6667)	15(23.3333)	50
Total	80	70	150

The χ^2 test gives

$$\chi_c^2 = \frac{20^2}{26.6667} + \frac{25^2}{26.6667} + \ldots + \frac{15^2}{23.3333} - 150$$
$$= 159.3752 - 150 = 9.3752.$$

This is compared to a $\chi^2((3-1)(2-1)) = \chi^2(2)$ and has associated p−value < 0.01. It is concluded that administration of Paraquat with agent A does affect the survival rate of rats.

Chapter 9

Single Factor Studies: Comparing Means and Determining Sample Sizes

9.1. (a) The linear combination $\widehat{\theta}_1$ is a contrast by the fact that $c_1 + c_2 + \ldots + c_t = 0$. To prove that $\widehat{\theta}_2$ is a contrast we require that $dc_1 + dc_2 + \ldots + dc_t = 0$. This is easily verified since $dc_1 + dc_2 + \ldots + dc_t = d(c_1 + c_2 + \ldots + c_t) = d(0) = 0$.

(b) In essence we need to verify that

$$\frac{|\widehat{\theta}_1|}{\widehat{\sigma}_{\widehat{\theta}_1}} = \frac{|\widehat{\theta}_2|}{\widehat{\sigma}_{\widehat{\theta}_2}}.$$

This can be shown since for any non-zero constant d,

$$|\widehat{\theta}_2| = |d||\widehat{\theta}_1|$$

and

$$
\begin{aligned}
\widehat{\sigma}_{\widehat{\theta}_2} &= \sqrt{\left(\frac{d^2 c_1^2}{n_1} + \frac{d^2 c_2^2}{n_2} + \ldots + \frac{d^2 c_t^2}{n_t}\right) MS[E]} \\
&= |d|\sqrt{\left(\frac{c_1^2}{n_1} + \frac{c_2^2}{n_2} + \ldots + \frac{c_t^2}{n_t}\right) MS[E]} \\
&= |d|\widehat{\sigma}_{\widehat{\theta}_1}.
\end{aligned}
$$

(c) Since the test statistic remains the same for both contrasts, this implies that the associated p-value will also remain the same. Since this is true for all non-zero d, this implies in general that multiplying a contrast by any non-zero constant will have no effect on the p-value.

(d) The test of significance is a test based on whether or not the contrast is zero. Testing H_0: $contrast = 0$ is the same as testing H_0: $a \times contrast = 0$.

9.3. (a) The parameter $\theta = \mu_2 - \mu_3$ can be written as $\theta = (0)\mu_1 + (1)\mu_2 + (-1)\mu_3$ so that $c_1 = 0, c_2 = 1$, and $c_3 = -1$. The unbiased estimator is

$$
\begin{aligned}
\hat{\theta} &= (0)\overline{Y}_1 + (1)\overline{Y}_2 + (-1)\overline{Y}_3 \\
&= \overline{Y}_2 - \overline{Y}_3 \\
&= 1.7475 - 1.5775 = 0.17.
\end{aligned}
$$

In Exercise **8.10** the MS[E] was calculated as 0.0057. This gives

$$\sigma_{\hat{\theta}} = \sqrt{\left[\left(\frac{0^2}{n_1} + \frac{1^2}{n_2} + \frac{(-1)^2}{n_3}\right) MSE\right]}$$

$$= \sqrt{\left[\left(\frac{1}{4} + \frac{1}{4}\right) 0.0057\right]}$$

$$= 0.1688.$$

Finally, $t(9, .025) = 2.262$. Then, a 95% confidence interval for θ is

$$\hat{\theta} \pm t(9, .025)\sigma_{\hat{\theta}}$$

or

$$.17 \pm 2.262(0.1688) = (-0.2118, 0.5518).$$

Since the interval does contain 0 there is not sufficient evidence to say that there is a difference in the mean soil bulk densities when 2 weeks of grazing is followed by 1 and 2 weeks of rest periods.

(b)   i. $\hat{\theta}_1$ estimates the difference between true mean bulk densities for continuous grazing and the mean of the true mean bulk densities for 2 wk grazing with 1 and 2 wk rest periods.

ii. $\hat{\theta}_2$ estimates the difference between the true mean bulk density for continuous grazing and the true mean bulk density for 2 wk grazing with 1 wk rest periods.

iii. $\hat{\theta}_3$ estimates the difference between the true mean bulk density for 2 wk grazing with 1 wk rest period and the true mean bulk density for 2 wk grazing with 2 wk rest period.

iv. $\hat{\theta}_4$ estimates the difference between the mean of the true mean bulk densities for continuous grazing and 2 wk grazing with 1 wk rest period and the true mean bulk density for 2 wk grazing period with 2 wk rest period.

(c) Because there are three treatments, there are two treatment degrees of freedom. Thus, there can only be two mutually orthogonal contrasts at a time.

(d) The two sets are $(\hat{\theta}_1, \hat{\theta}_3)$ and $(\hat{\theta}_2, \hat{\theta}_4)$. To see that they are orthogonal within the set, consider the first pair,

$$(\hat{\theta}_1, \hat{\theta}_3) = (\overline{Y}_1 - \frac{1}{2}(\overline{Y}_2 + \overline{Y}_3), \overline{Y}_2 - \overline{Y}_3)$$

so that $c_1 = 1$, $c_2 = -\frac{1}{2}$, $c_3 = -\frac{1}{2}$ and $d_1 = 0$, $d_2 = 1$, $d_3 = -1$. Then, we have

$$c_1 d_1 + c_2 d_2 + c_3 d_3 = (1)(0) + (-\frac{1}{2})(1) + (-\frac{1}{2})(-1) = 0.$$

Thus, these two contrasts are orthogonal. The other pair can be done similarly.

The first set would give the more meaningful subdivision. $\hat{\theta}_1$ can be interpreted as the difference in the effect on soil bulk density of continuous grazing versus grazing with a rest period. $\hat{\theta}_3$ can be interpreted as the difference in the effect on soil bulk density of 2 wk grazing with 1 wk rest versus 2 wks rest.

(e) From Exercise **8.4** the treatment means are $\overline{Y}_1 = 1.8925$, $\overline{Y}_2 = 1.7475$, and

$\overline{Y}_3 = 1.5775$. Then,

$$\hat{\theta}_1 = \overline{Y}_1 - \frac{1}{2}(\overline{Y}_2 + \overline{Y}_3) = 1.8925 - \frac{1}{2}(1.7475 + 1.5775) = 0.2300,$$

$$\hat{\theta}_3 = \overline{Y}_2 - \overline{Y}_3 = 1.7475 - 1.5775 = 0.1700$$

$$SS[\hat{\theta}_1] = \frac{\hat{\theta}_1^2}{\left(\frac{c_1^2}{n_1} + \frac{c_2^2}{n_2} + \frac{c_3^2}{n_3}\right)} = \frac{0.23^2}{\left(\frac{1^2}{4} + \frac{(-\frac{1}{2})^2}{4} + \frac{(-\frac{1}{2})^2}{4}\right)} = \frac{0.529}{0.375} = 0.14107, \text{and}$$

$$SS[\hat{\theta}_3] = \frac{\hat{\theta}_1^2}{\left(\frac{c_1^2}{n_1} + \frac{c_2^2}{n_2} + \frac{c_3^2}{n_3}\right)} = \frac{0.17^2}{\left(\frac{0^2}{4} + \frac{1^2}{4} + \frac{(-1)^2}{4}\right)} = \frac{0.0289}{0.5} = 0.0578$$

To test the significance of these contrasts we perform the following hypothesis.

H$_0$: $\theta_1 = \mu - \frac{1}{2}(\mu_2 + \mu_3) = 0$

H$_1$: $\theta_1 \neq 0$

Test Statistic: $F_c = \dfrac{MS[\hat{\theta}_1]}{MS[E]} = \dfrac{0.1411}{0.0057} = 24.7491$

$p-$value $< .025$

Conclusion: Reject H$_0$.

H$_0$: $\theta_3 = \mu_2 - \mu_3 = 0$

H$_1$: $\theta_3 \neq 0$

Test Statistic: $F_c = \dfrac{MS[\hat{\theta}_3]}{MS[E]} = \dfrac{0.0578}{0.0057} = 10.1403$

$p-$value $< .025$

Conclusion: Reject H$_0$.

There is sufficient evidence ($p-$value $<.025$) to indicate that the mean soil bulk density for continuous grazing is different than the mean soil bulk density for grazing with rest periods and that the mean soil bulk densities are different for grazing with 1 and 2 wk rest periods.

9.5. Let us denote the means in the table as follows:

P–Release	Inoculation	Foliar P	
		No	Yes
LOW	NO	$\overline{Y}_1$	$\overline{Y}_7$
	YES	$\overline{Y}_2$	$\overline{Y}_8$
MEDIUM	NO	$\overline{Y}_3$	$\overline{Y}_9$
	YES	$\overline{Y}_4$	$\overline{Y}_{10}$
HIGH	NO	$\overline{Y}_5$	$\overline{Y}_{11}$
	YES	$\overline{Y}_6$	$\overline{Y}_{12}.$

In the above table, $\overline{Y}_i$ is an estimate of the population mean μ_i.

(a)   i. $\theta_1 = \frac{1}{6}(\mu_1 + \mu_2 + \mu_3 + \ldots + \mu_6) - \frac{1}{6}(\mu_7 + \mu_8 + \ldots + \mu_{12})$.

   ii. $\theta_2 = \frac{1}{6}(\mu_1 + \mu_7 + \mu_3 + \mu_9 + \mu_5 + \mu_{11}) - \frac{1}{6}(\mu_2 + \mu_8 + \mu_4 + \mu_{10} + \mu_6 + \mu_{12})$.

   iii. $\theta_3 = \frac{1}{4}(\mu_1 + \mu_2 + \mu_7 + \mu_8) - \frac{1}{4}(\mu_5 + \mu_6 + \mu_{11} + \mu_{12})$.

   iv. $\theta_4 = \frac{1}{2}(\mu_5 + \mu_{11}) - \frac{1}{2}(\mu_6 + \mu_{12})$.

As noted in Exercise **9.1**, each of the above contrasts can be multiplied by a constant without altering the results of the hypothesis tests. For example, the hypothesis tests for $\theta_1^* = 6\theta_1 = (\mu_1 + \ldots + \mu_6) - (\mu_7 + \ldots + \mu_{12})$ would yield the same result as that for θ_1. It is left to the reader to verify that the four contrasts are orthogonal.

(b) Using Exercise **8.13** we have

$$MS[E] = .5711 \quad \text{and} \quad n_i = 3 \text{ for all } i.$$

The standard errors of the contrasts can then be found as

$$
\begin{aligned}
\hat{\sigma}_{\hat{\theta}_1} &= \sqrt{\left(\frac{(1/6)^2}{3} + \cdots + \frac{(1/6)^2}{3} + \frac{(-1/6)^2}{3} + \cdots + \frac{(-1/6)^2}{3}\right) MS[E]} \\
&= \sqrt{12\left(\frac{1}{108}\right).5711} = 0.2519. \\
\hat{\sigma}_{\hat{\theta}_2} &= \sqrt{\left(\frac{(1/6)^2}{3} + \frac{(-1/6)^2}{3} + \cdots + \frac{(1/6)^2}{3} + \frac{(-1/6)^2}{3}\right) MS[E]} \\
&= \sqrt{12\left(\frac{1}{108}\right).5711} = 0.2519. \\
\hat{\sigma}_{\hat{\theta}_3} &= \sqrt{\left(\frac{(1/4)^2}{3} + \frac{(1/4)^2}{3} + \frac{(-1/4)^2}{3} + \frac{(-1/4)^2}{3} + \cdots\right) MS[E]} \\
&= \sqrt{8\left(\frac{1}{48}\right).5711} = 0.3085. \\
\hat{\sigma}_{\hat{\theta}_4} &= \sqrt{\left(\frac{(1/2)^2}{3} + \frac{(-1/2)^2}{3} + \frac{(1/2)^2}{3} + \frac{(-1/2)^2}{3}\right) MS[E]} \\
&= \sqrt{4\left(\frac{1}{12}\right).5711} = 0.4363
\end{aligned}
$$

The $\hat{\theta}_i$'s can be found by substituting the sample means in the above contrast statements. Doing so leads to the following estimates:

$$\hat{\theta}_1 = .0833, \quad \hat{\theta}_2 = -1.8167, \quad \hat{\theta}_3 = -2.925, \quad \hat{\theta}_4 = -1.45.$$

The individual confidence intervals all have the same form

$$\hat{\theta}_i \pm t(N - t, .025)\hat{\sigma}_{\hat{\theta}_i}.$$

For θ_1 the confidence interval is

$$.0833 \pm t(24, .025)(.2519) = .0833 \pm 2.064(.2519) \quad \text{or} \quad (-0.4366, 0.6032).$$

Since this confidence interval contains zero, the null hypothesis is in the range of plausible values and hence, at the 0.05 significance level we conclude that there is insufficient evidence to believe that there is a difference, in mean leaf area ratios, due to the variable Foliar P when averaged over all other factors.

For θ_2 the confidence interval is

$$-1.8167 \pm t(24, .025)(.2519) = -1.8167 \pm 2.064(.2519) \quad \text{or} \quad (-2.3366, -1.2968).$$

Since this interval does not contain zero, the null hypothesis is not in the range of plausible values and hence, we reject the null hypothesis. That is, at the 0.05 significance level we conclude that there is a difference between the true mean leaf area ratios for treatments with and without inoculation. In fact there seems to be strong evidence that the true mean leaf area ratio for inoculated plants is larger than for those without inoculation.

For θ_3 the confidence interval is

$$-2.925 \pm t(24, .025)(.3085) = -2.925 \pm 2.064(.3085) \quad \text{or} \quad (-3.5617, -2.2883).$$

Since this interval does not contain zero, the null hypothesis is not in the range of plausible values and hence, we reject the null hypothesis. That is, at the 0.05 significance level we conclude that there is a difference between the true mean leaf area ratios of the high and low release of P. In fact there seems to be strong evidence that the true mean leaf area ratio for plants treated with high P–Release is larger than with low P–Release.

For θ_4 the confidence interval is

$$-1.45 \pm t(24, .025)(.4363) = -1.45 \pm 2.064(.4363) \quad \text{or} \quad (-2.3505, -0.5495).$$

Since this interval does not contain zero, the null hypothesis is not in the range of plausible values and hence, we reject the null hypothesis. That is, at the 0.05 significance level we conclude that there is a difference between the true mean leaf area ratios of the high P-release treatments with and without inoculation. In fact there seems to be some evidence that the true mean leaf area ratio for inoculated high P–Release plants is larger than the uninoculated high P–Release plants.

9.7. Defining the four means of interest in the table as

P–Release	Inoculation	Foliar P	
		No	Yes
LOW	NO	$\overline{Y}_1$	$\overline{Y}_3$
	YES	$\overline{Y}_2$	$\overline{Y}_4$.

(a) The appropriate contrast would then take the form

$$\hat{\theta} = \left(\overline{Y}_3 - \overline{Y}_1\right) - \left(\overline{Y}_4 - \overline{Y}_2\right).$$

(b) Using the fact that $MS[E] = .5711$, as calculated in Exercise **8.13**, the standard error of the contrast is

$$\hat{\sigma}_{\hat{\theta}} = \sqrt{\left(\frac{(-1)^2}{3} + \frac{1^2}{3} + \frac{1^2}{3} + \frac{(-1)^2}{3}\right) MS[E]}$$

$$= \sqrt{\frac{4}{3}(.5711)} = .8726.$$

Since the contrast was tested after looking at the data, we use the Scheffe method and test $H_0 : \theta = 0$. The null hypothesis is rejected at $\alpha = 0.01$ if

$$|\hat{\theta}| > \sqrt{(t-1)F(t-1, N-t, .01)}\, \hat{\sigma}_{\hat{\theta}}$$

$$> \sqrt{11F(11, 24, .01)}\, .8726 = \sqrt{11(1.86)}\, .8726 = 3.947.$$

The estimated value of the contrast is $\hat{\theta} = (10.7 - 11.6) - (13.5 - 12.8) = -1.6$. Hence the null hypothesis is not rejected. There is insufficient evidence to conclude that at low P–Release, there is a difference between the effects of Foliar P application based on whether or not the plants have been inoculated.

9.9. (a) From Exercise **9.3** we have $\hat{\theta}_1 = 0.2300$ and $\hat{\theta}_3 = 0.1700$. From the text,

$$\hat{\sigma}_{\hat{\theta}_i} = \sqrt{\left[\frac{c_1^2}{n_1} + \frac{c_2^2}{n_2} + \frac{c_3^2}{n_3}\right] MS[E]}.$$

Now, the other estimates and estimated standard errors are

$$\hat{\sigma}_{\hat{\theta}_1} = \sqrt{\left[\frac{1^2}{4} + \frac{\left(\frac{-1}{2}\right)^2}{4} + \frac{\left(\frac{-1}{2}\right)^2}{4}\right] 0.0057} = 0.0462,$$

$$\hat{\sigma}_{\hat{\theta}_3} = \sqrt{0.5(0.0057)} = 0.0534,$$

$$\hat{\theta}_2 = \overline{Y}_1 - \overline{Y}_2 = 1.8925 - 1.7475 = 0.145,$$

$$\hat{\sigma}_{\hat{\theta}_2} = \sqrt{\left[\frac{(1)^2}{4} + \frac{(-1)^2}{4}\right] 0.0057} = 0.0534,$$

$$\hat{\theta}_4 = \frac{1}{2}(1.8925 + 1.7475) - 1.5775 = 0.2425,$$

and

$$\hat{\sigma}_{\hat{\theta}_4} = \sqrt{\left[\frac{\left(\frac{1}{2}\right)^2}{4} + \frac{\left(\frac{1}{2}\right)^2}{4} + \frac{(-1)^2}{4}\right] 0.0057} = 0.0462.$$

From these results, the Bonferroni comparisons are

$$\left|\hat{\theta}_i\right| > t(\nu, \alpha/2k)\hat{\sigma}_{\hat{\theta}_i}.$$

Then, with $\nu = 9$ and $k = 4$ in Table C.10, we get $t(9, .05/8) = 3.111$, so that

θ_i	$\hat{\theta}_i$	CCV	Conclusion
θ_1	0.2300	0.1437	Significant
θ_2	0.1450	0.1166	Significant
θ_3	0.1700	0.1166	Significant
θ_4	0.2425	0.1437	Significant

Hence, there is sufficient evidence at the 0.05 level to indicate that the four contrasts are simultaneously significantly different from zero.

The Scheffe comparison is

$$\left|\hat{\theta}_i\right| > \sqrt{(t-1)F(t-1, \nu, \alpha)}\,\hat{\sigma}_{\hat{\theta}_i}$$

with $F(2, 9, .05) = 4.2565$ and $\sqrt{(t-1)F(t-1, \nu, \alpha)} = 2.9177$. The summary of the information is

θ_i	$\hat{\theta}_i$	CCV	Conclusion
θ_1	0.2300	0.1348	Significant
θ_2	0.1450	0.1558	Not Significant
θ_3	0.1700	0.1558	Significant
θ_4	0.2425	0.1348	Significant

There is sufficient evidence, at the 0.05 level, to indicate that θ_1, θ_3, and θ_4 are significantly different from zero, but there is not sufficient evidence that θ_2 is.

(b) The Fisher method uses the comparison

$$\left|\hat{\theta}_i\right| > t(\nu, \alpha/2)\hat{\sigma}_{\hat{\theta}_i}$$

where $t(\nu, \alpha/2) = t(9, .025) = 2.262$. The comparisons are

θ_i	$\hat{\theta}_i$	CCV	Conclusion
θ_1	0.2300	0.1045	Significant
θ_3	0.1700	0.1208	Significant

The comparison here is equivalent to using the rejection region based on the t-statistic

$$t = \frac{\left|\hat{\theta}_i\right|}{\hat{\sigma}_{\hat{\theta}_i}} > t(\nu, \alpha/2).$$

It has been shown that the t-test is equivalent to the one degree of freedom F-test which was the test used in Exercise **9.3**. Therefore, we expect to get the same results.

(c) For these two comparisons, the Bonferroni t-value is $t(9, .05/4) = 2.6850$. The results are

θ_i	$\hat{\theta}_i$	CCV	Conclusion
θ_1	0.2300	0.1240	Significant
θ_3	0.1700	0.1434	Significant

For Scheffe's comparisons, $F(2, 9, .05) = 4.256$ and $\sqrt{(t-1)F(2, 9, .05)} = 2.9177$. We then have

θ_i	$\hat{\theta}_i$	LSD	Conclusion
θ_1	0.2300	0.1348	Significant
θ_3	0.1700	0.1558	Significant

The results for the Scheffe method are the same as those in Part Part **a** in Exercise **9.9** because the CCV for the Scheffe does not depend on the number of contrasts. Hence, the CCV is the same for two and four contrasts.

9.11. From Exercise **8.3**, and denoting the Low treatment by 1, Medium treatment by 2 and High treatment by 3, we have

$$\overline{Y}_1 = 2.2325, \quad \overline{Y}_2 = 3.4375, \quad \overline{Y}_3 = 4.50,$$

$$MS[E] = \frac{S_1^2 + S_2^2 + S_3^2}{3} = \frac{.9051 + .2121 + .5426}{3} = .5533, \quad N = 12, \quad t = 3, \quad n_i = n = 4.$$

The multiple comparisons procedures can be broken into two cases. Those procedures which depend on how many other differences fall within their range (Student–Neuman–Keuls and Duncan) and those procedures that compare all differences to the same value. **These tests will all be tested using $\alpha = .05$.**

The Bonferroni method compares each difference to

$$t(N - t, \tfrac{\alpha}{t(t-1)})\sqrt{\frac{2}{n}MS[E]} = t(9, .0083)\sqrt{\frac{2}{4}(.5533)} = 1.5442,$$

where $t(9, .0083)$ corresponds to taking $\nu = 9$ and $k = 6$ in the reference table. The Scheffe method compares each difference to

$$\sqrt{(t-1)F(t-1, N-t, \alpha)\left(\frac{2}{n}\right)MS[E]} = \sqrt{2F(2, 9, .05)\left(\frac{2}{4}\right).5533} = 1.5346.$$

The Tukey method compares each difference to

$$q(t, N - t, \alpha)\sqrt{\frac{MS[E]}{n}} = q(3, 9, .05)\sqrt{\frac{.5533}{4}} = 1.4691.$$

81

The Student–Neuman–Keuls method has each difference depend on

$$q(p, N-t, \alpha)\sqrt{\frac{MS[E]}{n}} = q(p, 9, .05)\sqrt{\frac{.5533}{4}}.$$

The Duncan procedure has each difference depend on

$$R(p, N-t, \alpha)\sqrt{\frac{MS[E]}{n}} = R(p, 9, .05)\sqrt{\frac{.5533}{4}}.$$

Here p has the same interpretation as above.

The following table will test all pairs of means.

Difference	$\overline{Y}_2 - \overline{Y}_1$ 1.205	$\overline{Y}_3 - \overline{Y}_1$ 2.2675	$\overline{Y}_3 - \overline{Y}_2$ 1.0625
p	2	3	2
Bonferroni	$1.5442(NS)$	$1.5442(S)$	$1.5442(NS)$
Scheffe	$1.5346(NS)$	$1.5346(S)$	$1.5346(NS)$
Tukey	$1.4691(NS)$	$1.4691(S)$	$1.4691(NS)$
Student–Neuman–Keuls	$1.1901(S)$	$1.4691(S)$	$1.1901(NS)$
Duncan	$1.1901(S)$	$1.2422(S)$	$1.1901(NS)$

Based on the above tests, there is strong evidence, at the $\alpha = .05$ level that treatments 1 and 3 are significantly different. There is some evidence that treatments 1 and 2 are also different (based on Student–Neuman–Keuls and Duncan). The most powerful test for all comparisons is the Duncan test since it has the smallest value for the LSD. The first three tests all constrain the experiment-wise error rate to be .05. The other two do not constrain the experimental error rate.

9.13. Using the results of the TRCP data in Exercise 8.11 we get

$$\overline{Y}_1 = 4.642 \quad \overline{Y}_2 = 3.575 \quad \overline{Y}_3 = 3.3071,$$

$$MS[E] = .2228, \quad N = 16, \quad t = 3, \quad n_1 = 5, \quad n_2 = 4, \quad n_3 = 7.$$

All tests will be at the $\alpha = .05$ level. The first thing to notice here is that the sample sizes are unequal. Because of this, harmonic means will be used for the calculations. For comparing treatments 1 and 2 the harmonic mean denoted n_{12} is

$$n_{12} = \frac{2}{\frac{1}{n_1} + \frac{1}{n_2}} = \frac{2}{\frac{1}{5} + \frac{1}{4}} = 4.4444.$$

For comparing treatments 1 and 3 the harmonic mean denoted n_{13} is

$$n_{13} = \frac{2}{\frac{1}{n_1} + \frac{1}{n_3}} = \frac{2}{\frac{1}{5} + \frac{1}{7}} = 5.8333.$$

For comparing treatments 2 and 3 the harmonic mean denoted n_{23} is

$$n_{23} = \frac{2}{\frac{1}{n_2} + \frac{1}{n_3}} = \frac{2}{\frac{1}{4} + \frac{1}{7}} = 5.0909.$$

The Student–Neuman–Keuls method has the difference between means i and j depend on

$$q(p, N-t, \alpha)\sqrt{\frac{MS[E]}{n_{ij}}} = q(p, 13, .05)\sqrt{\frac{.2228}{n_{ij}}}.$$

Here p is the number of means whose values fall between the two means being compared. The Duncan procedure has the difference between means i and j depend on

$$R(p, N-t, \alpha)\sqrt{\frac{MS[E]}{n_{ij}}} = R(p, 13, .05)\sqrt{\frac{.2228}{n_{ij}}}.$$

Here p has the same interpretation as above. The following table will test all pairs of means.

Difference	$\overline{Y}_1 - \overline{Y}_2$	$\overline{Y}_1 - \overline{Y}_3$	$\overline{Y}_2 - \overline{Y}_3$
	1.067	1.3349	.2679
n_{ij}	4.4444	5.8333	5.0909
p	2	3	2
SNK	.6851(S)	.7290(S)	.6401(NS)
Duncan	.6851(S)	.6273(S)	.6401(NS)

Based on the above tests there seems go be strong evidence that treatments 1 and 2 differ and that 1 and 3 differ. Both tests showed these as significantly different. In terms of power, the Duncan test is more powerful for all comparisons. This can be seen because the value that the differences are compared to is smaller than the values for the Student–Neuman–Keuls procedure.

9.15. The four tests in order of conservativeness are

Most Conservative	Scheffe
	Tukey
	Duncan
Least Conservative	Fisher

Since item a) shows the fewest treatment differences this must correspond to Scheffe's procedure. Since item d) has the next fewest treatment differences, this must correspond to Tukey's procedure. Since item b) shows the most treatment differences this must correspond to the Fisher procedure. Item c) is the only one left, so this must correspond to Duncan's procedure.

9.17. (a) The 90% Bonferroni simultaneous confidence intervals for the set of four contrasts given in Exercise **9.3** are

$$\hat{\theta}_i \pm t(\nu, \alpha/2k)\sigma_{\hat{\theta}_i}$$

where $t(9, 0.10/2(4)) = t(9, .10/8) = t(9, .0125) = 2.685$. With this information, the specific intervals are

θ_1: $0.2300 \pm 2.685(0.0462) = (0.1059, 0.3540),$

θ_2: $0.1450 \pm 2.685(0.0534) = (0.0016, 0.2884),$

θ_3: $0.1700 \pm 2.685(0.0534) = (0.0266, 0.3134),$ and

θ_4: $0.2425 \pm 2.685(0.0462) = (0.1184, 0.3665).$

These confidence intervals indicate that, at the overall 90% confidence level, all four contrasts are simultaneously statistically different from zero.

(b) For the two meaningful contrasts the 90% Bonferroni simultaneous confidence intervals are calculated in the same way, but with

$$t(9, 0.10/4) = t(9, 0.025) = 2.2622.$$

Then, the confidence intervals are

θ_1: $0.2300 \pm 2.2622(0.0462) = (0.1255, 0.3345),$ and

θ_3: $0.1700 \pm 2.2622(0.0534) = (0.0492, 0.2908)$.

These intervals are not as wide as their counterparts in part Part **a** in Exercise **9.17**, because each interval is calculated at a lower confidence level. The difference in the conclusions is due to the fewer comparisons here than in part Part **a** in Exercise **9.17**. The Bonferroni procedure controls the experiment-wise error rate. Dividing the error rate among more comparisons reduces the error rate for each comparison and, thus, requires greater confidence. The confidence intervals, in turn, are wider.

(c) The set of 95% Tukey simultaneous confidence intervals for all possible pairwise differences between the three population means are calculated as

$$\overline{Y}_i - \overline{Y}_k \pm q(t, \nu, \alpha) \frac{1}{\sqrt{2}} \hat{\sigma}_{\overline{Y}_i - \overline{Y}_k}$$

for some i and k. From Table C.11 $q(3, 9, 0.05) = 3.95$. In Exercise **8.10** MS[E] was calculated as 0.0057. This gives

$$\hat{\sigma}_{\hat{\theta}_{\overline{Y}_i - \overline{Y}_k}} = \sqrt{\frac{2\mathrm{MS[E]}}{n}} = \sqrt{\frac{2(0.0057)}{4}} = 0.0534$$

and then the margin of error is

$$q(3, 9, 0.05) \frac{1}{\sqrt{2}} \hat{\sigma}_{\overline{Y}_i - \overline{Y}_k} = (3.95)(0.7071)(0.0534) = 0.1493.$$

The treatment means were calculated in Exercise **8.4**. Then the intervals are

$\mu_1 - \mu_2$: $1.8925 - 1.7475 \pm 0.1493 \rightarrow (-0.0043, 0.2943)$
$\mu_2 - \mu_3$: $1.7475 - 1.5775 \pm 0.1493 \rightarrow (0.0207, 0.3193)$
$\mu_1 - \mu_3$: $1.8925 - 1.5775 \pm 0.1493 \rightarrow (0.1657, 0.4643)$

These intervals indicate that at the overall 0.05 significance level population mean of the third treatment is significantly different from the population means of the first and second treatments, but we can't say that the population means of the first and second treatments are different.

(d) The Tukey and Bonferroni procedures each use a different coefficient. Tukey uses the studentized range times $1/\sqrt{2}$, while Bonferroni uses the t-distribution. This causes the Tukey intervals to be wider than the Bonferroni and, thus, more conservative.

9.19. The easiest way to compare the two intervals is to look at their respective widths. Taking the upper confidence limit minus the lower confidence limit gives the width of the interval. For the Bonferroni simultaneous intervals the width is

$$2t(30, \tfrac{.05}{20})\hat{\sigma}_{\hat{\theta}_i}.$$

For the Tukey procedure the width of the intervals is

$$2q(10, 30, .05)\hat{\sigma}_{\hat{\theta}_i}.$$

Thus comparing the widths is equivalent to comparing $t(30, \tfrac{.05}{20})$ to $q(10, 30, .05)$. Using the tables in the text

$$t(30, \tfrac{.05}{20}) = 3.0298$$
$$q(10, 30, .05) = 4.82$$

This implies that the simultaneous 95% confidence intervals based on the Bonferroni procedure will be narrower.

9.21. (a) Box 9.10 gives the 95% individual prediction intervals as

$$\hat{\theta} \pm t(N - t, \alpha/2)\sigma_{\hat{\theta}_f}.$$

Since we are calculating prediction intervals about a single value, i.e. $m = 1$, we have

$$\hat{\theta}_f = \overline{Y}_{i,f} \text{ where } i = SC, SW, FC, \text{ or } FW$$

and

$$\sigma_{\hat{\theta}_f} = \sqrt{\left(1 + \frac{1}{n}\right)MS[E]} = \sqrt{\left(1 + \frac{1}{3}\right)0.0043} = 0.2394.$$

With $t(8, 0.025)$, the 95% individual prediction intervals are

Condition	$\hat{\theta}$	Interval
SC	1.55	$(0.998, 2.102)$
SW	1.59	$(1.038, 2.142)$
FC	1.08	$(0.528, 1.632)$
FW	2.01	$(1.458, 2.562)$

The first interval can be interpreted as with 95% confidence the future weight gain for a fish in still, cold water is between 0.998 and 2.102 lbs. The other intervals are interpreted similarly.

(b) To construct a set of 95% simultaneous prediction intervals we use the same values as in Part a in Exercise **9.21**, but with the Bonferroni method. That is, we use $t(8, 0.05/24) = 3.3554$. Hence, the intervals are

Condition	$\hat{\theta}$	Interval
SC	1.55	$(0.7467, 2.3533)$
SW	1.59	$(0.7867, 2.3933)$
FC	1.08	$(0.2767, 1.8833)$
FW	2.01	$(1.2067, 2.8133)$

With 95% confidence the future weight gain of fish under all four conditions will be within their respective intervals.

9.23. (a) From the text we have

$$\mu_1 = 1.8, \ \mu_2 = 1.7, \ \mu_3 = 1.5, \ 1 - \beta = 0.80, \ \alpha = 0.05.$$

Using these values we have

$$\sigma_\mu^2 = \frac{1}{3}\sum(\mu_i - \overline{\mu})^2 = 0.0156.$$

The formula for λ is

$$\lambda = \frac{nt\sigma_\mu^2}{2\sigma^2}$$

where σ^2 can be estimated by $MS[E] = 0.0057$ (as in Exercise **8.10**. Using S-plus we arrive at the following table

n	ν_1	ν_2	λ	$1 - \beta$
2	2	3	8.1871	0.5507
3	2	6	12.2807	0.9278
4	2	9	16.3743	0.9927

Therefore, collecting three replications per treatment will give the desired power level.

(b) From the text we have

$$\Delta = 0.1 \text{ and } 1 - \beta = 0.85.$$

As mentioned in the text and argued in Exercise **9.25**, using

$$\sigma_\mu^2 = \frac{\Delta^2}{2t} = \frac{0.1^2}{2(3)} = 0.00167$$

to calculate the non-centrality parameter will ensure a power of at least $1 - \beta$ for detecting a minimum difference of Δ between at least one pair of means. Assuming $\alpha = 0.05$ gives the following table.

n	ν_1	ν_2	λ	$1 - \beta$
9	2	24	3.9474	0.6525
10	2	27	4.3860	0.7081
11	2	30	4.8246	0.7566
12	2	33	5.2632	0.7984
13	2	36	5.7018	0.8341
14	2	39	6.1404	0.8643
15	2	42	6.5789	0.8896

Therefore, taking fourteen replications per treatment will give the desired power level.

(c) From the text we have that $1 - \beta = 0.90$ and we can assume a given α of 0.05, say. The text says that "the probability is 90% (or more) that the amount of insulin released at one selected level exceeds the amount released at the other level." Another way to say this is that the probability is at least 90% that the difference between the amount of insulin released at two selected levels is greater than zero.

Since we are assuming that the errors are approximately normally distributed, we can assume that the difference between two amounts, $Y_{i1} - Y_{j1}$ for $i \neq j$ is approximately normally distributed with mean $\mu_i - \mu_j$ and variance $2\sigma^2$. Hence, the above requirement can be estimated as

$$Pr\{Y_{i1} - Y_{j1} > 0\} \geq 0.90$$

$$\Rightarrow Pr\left\{\frac{(Y_{i1} - Y_{j1}) - (\mu_i - \mu_j)}{\sigma\sqrt{2}} > \frac{0 - (\mu_i - \mu_j)}{\sigma^2\sqrt{2}}\right\} \geq 0.90$$

$$\Rightarrow Pr\left\{Z > \frac{-\Delta}{\sigma\sqrt{2}}\right\} \geq 0.90$$

where $Z \sim N(0,1)$. The value for $\frac{-\Delta}{\sigma\sqrt{2}}$ that makes the last inequality an equality is -1.2815. Now we solve

$$\frac{-\Delta}{\sigma\sqrt{2}} = -1.2815$$

$$\Rightarrow \frac{\Delta}{\sigma} = 1.8123.$$

Therefore, the requirement given in the text is equivalent to the minimum distance between two means over the standard deviation exceeding 1.8123. Now, using the S-plus statistical software we arrive at the following table.

n	ν_1	ν_2	λ	$1 - \beta$
5	2	12	4.1087	0.6110
6	2	15	4.9305	0.7193
7	2	18	5.7522	0.8025
8	2	21	6.5739	0.8640
9	2	24	7.3957	0.9081
10	2	27	8.2174	0.9390

Therefore, using nine replications will give the desired level of power.

(d) This part considers sample sizes required for a desired accuracy of estimation. The text gives

$$B = 0.1, \; \alpha = 0.10; \; \theta_1 = \mu_1 - \mu_2; \; \theta_2 = \mu_1 + 2\mu_2 - \mu_3.$$

Because we are not looking at all possible contrasts or contrasts suggested by the data (Scheffe) nor at all possible pairwise comparisons, the two contrasts can be compared using the Bonferroni method. Then we also have

$$C_1 = (1)^2 + (-1)^2 = 2, \; C_2 = (1)^2 + (2)^2 + (-1)^2 = 6, \; \text{and } C = \max(2, 6) = 6.$$

Because we are using the Bonferroni method, the value for T is $t(\nu, \alpha, k) = t(3(r-1), 0.10, 2)$. Then, using a statistical package the following table can be considered.

n	ν	T	RHS
2	3	3.1824	11.5459
3	6	2.4469	6.8256
4	9	2.2621	5.8338
5	12	2.1788	5.4118
6	15	2.1314	5.1791
7	18	2.1009	5.0318

From the table, the first value of n which is greater than the right hand side is $n = 6$. Therefore, six replications per treatment will give the desired power.

(e) To make all possible pairwise comparisons, Tukey's method will be used. For this method $C = 2$, since all contrasts are the same. Then, from the text we have $\alpha = 0.05$ and $B = 0.5$. This gives the following table

n	ν	T	RHS
2	3	5.9097	9.9534
3	6	4.3392	5.3662
4	9	3.9485	4.4433
5	12	3.7729	4.0570
6	15	3.6734	3.8457

From the table, the first value of n which is greater than the right side is $n = 5$. Therefore, five replications per treatment will give the desired power.

9.25. From the comment in the text, when $|\mu_i - \mu_j| \geq \Delta$, the smallest value that σ_μ^2 can take on is $\Delta^2/(2t)$ and, therefore, the power, $1 - \beta$, calculated with this value will be the smallest power achievable with a particular sample size. Since the power increases with σ_μ^2 any other value of σ_μ^2 greater than $\Delta^2/(2t)$ will have a corresponding power greater than the above minimum. Therefore, using $\Delta^2/(2t)$ will ensure a power of at least $1 - \beta$ for detecting a difference Δ between at least one pair of treatments.

Chapter 10

Simple Linear Regression

10.1. (a) The one-way ANOVA model can be written as

$$Y_{ij} = \mu + \alpha_i + E_{ij}, \qquad i = 1, 2, 3, 4; j = 1, 2, 3, 4$$

where

Y_{ij} = measured chemical influx of jth rat given i^{th} dose,

μ = overall mean chemical influx,

α_i = effect of ith dose,

E_{ij} = random error.

(b) Dependent Variable: INFLUX

Source	DF	Sum of Squares	Mean Square	F Value	Pr > F
Model	3	2152.2500	717.4167	160.92	0.0001
Error	12	53.5000	4.4583		
Corrected Total	15	2205.7500			

The p−value, for testing $H_0 \mu_1 = \mu_2 = \mu_3 = \mu_4$, in the ANOVA is much smaller than $\alpha = 0.05$. Therefore, we would conclude that there is sufficient evidence to indicate that the mean response for the different treatments are statistically significantly different from each other. This implies that drug dose is affecting the chemical influx.

(c) A 95% Prediction Interval for the chemical influx in the brain of a rat receiving a dose of $d = 100$ nM of the experimental drug would be based on the formula

$$\overline{Y}_3 \pm t(\text{error df}, \alpha/2)\sqrt{MS[E](1 + \frac{1}{n_3})}.$$

The standard error is

$$\sqrt{4.4583(1 + \frac{1}{4})} = 2.3607.$$

Since $t(12, 0.025) = 2.179$ this would be

$$46.00 \pm 2.179(2.3607)$$

which is $(40.8560, 51.1440)$. Thus, with 95% confidence the chemical influx in the brain of a rat receiving a dose of $d = 100$ nM of the experimental drug is between 40.8560 and 51.1440 nmol/mg protein.

(d) A 95% lower confidence bound for the change in mean chemical influx when drug dose is changed from 1 to 10 nM is based on the formula

$$\overline{Y}_2 - \overline{Y}_1 - t(\text{error df}, \alpha/2)\sqrt{MS[E]\left(\frac{1}{n_2} + \frac{1}{n_1}\right)}.$$

Then the actual interval becomes

$$36.25 - 24 - 2.179\sqrt{4.4583\left(\frac{1}{4} + \frac{1}{4}\right)}$$

which is 8.9967. Thus, with 95% confidence we conclude that the mean chemical influx when drug dose is changed from 1 to 10 nM increases by at least 8.9967 nmol/mg protein.

10.3. (a) The figure below shows the scatterplot of Weight Gain versus Mother's Age. The weight gains rise linearly with age.

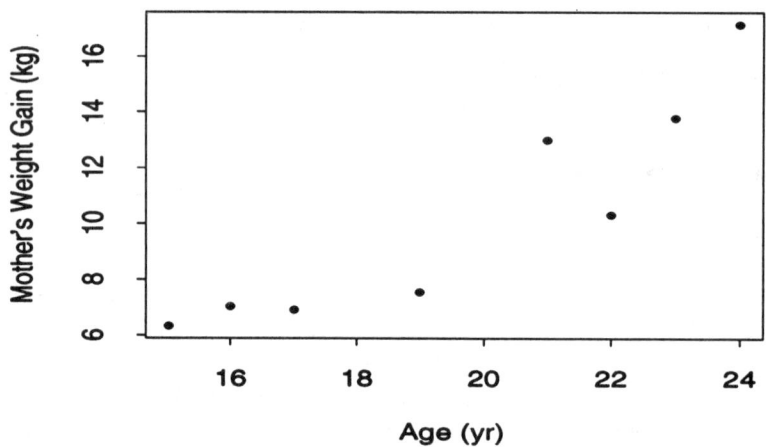

(b) Yes. With eight levels, or treatments, of Age there is no replication. Since there is no replication within a treatment there would be no estimate of experimental error. Without an estimate of experimental error the investigator could not perform any hypothesis tests or calculate any confidence intervals. This is a very serious drawback to the investigator's analysis.

10.5. (a) Once again, denoting Soil Water Content by Y and Depth by X, a simple linear regression model is

$$Y_i = \beta_0 + \beta_1 x_i + E_i \qquad i = 1, 2, 3, \ldots, 16$$

where

Y_i = the ith measured Soil Water Content,

β_0 = the mean Soil Water Content when Depth is zero,

β_1 = is the expected change in Soil Water Content for a unit increase in Depth,

E_i = random error independently distributed as $N(0, \sigma^2)$.

(b) The ANOVA model assumes that the mean of Soil Water Content might be different for each level of depth and there is no particular relationship between these means at different levels. The simple linear regression approach assumes that the mean of Soil

Water Content is different at each level of Depth, but the mean changes linearly with depth. In other words, the difference in means of Soil Water Content for any two equally spaced Depths is the same no matter what the Depth is.

(c) For all the means to be the same in the regression model, β_1 would have to be zero. Hence the null hypothesis is H_0:

$$\beta_1 = 0.$$

(d) Looking at the plot once again:

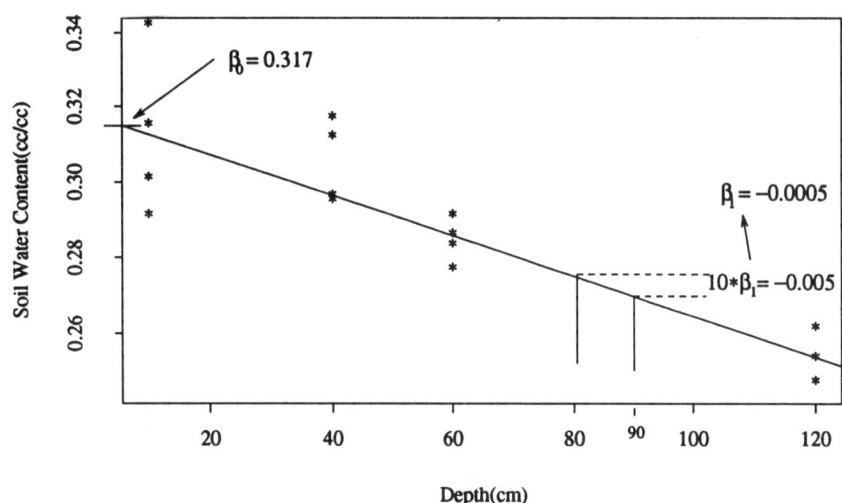

It can be seen that an appropriate line is of the form

$$\hat{y} = 0.317 - 0.0005x.$$

(e) It seems that the soil depth is held fixed by the experimenter because only four distinct levels of the independent variable and exactly four observations at each.

10.7. (a) Denoting Disease Index as Y and Crop Interval as X, a simple linear regression model is

$$Y_i == \beta_0 + \beta_1 x_i + E_i \qquad i = 1, \ldots, 10$$

where

Y_i = the ith measured level of Disease Index,
β_0 = is the mean Disease Index when Crop Interval is zero,
β_1 = is the expected change in Disease Index for a unit increase in Crop Interval,
E_i = random error independently distributed $N(0, \sigma^2)$.

(b) A plot of Disease Index vs. Crop Interval has the form:

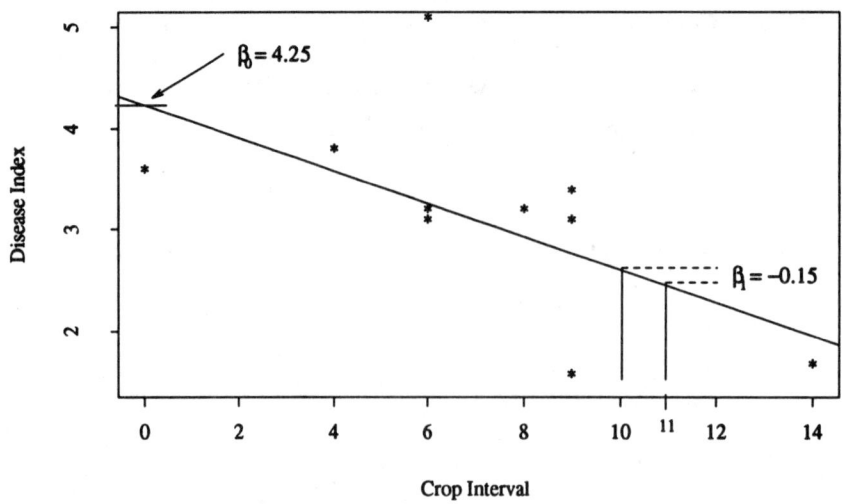

It can be seen that an appropriate line is of the form

$$\hat{y} = 4.25 - .15x.$$

(c) It seems unreasonable that the experimenter would be able to control the number of years that a crop could be harvested without peas or other legumes being planted. After all, one crop hadn't had peas planted on it for 14 years. This indicates that the Crop Interval was not held fixed by the investigator.

10.9. (a) The summary statistics for the chemical influx data with $Y = $ Chemical influx and $X = \log(d)$ are

$$\bar{x} = 3.4550$$
$$\bar{y} = 40.3750$$
$$S_{xx} = 106.1659$$
$$S_{yy} = 2205.7436$$
$$S_{xy} = 476.9200$$

The least squares estimates are

$$\hat{\beta}_1 = \frac{S_{xy}}{S_{xx}} = \frac{476.9200}{106.1659} = 4.4922,$$
$$\hat{\beta}_0 = \bar{y} - \hat{\beta}_1\bar{x} = 40.3750 - 4.4922(3.4550) = 24.8544.$$

Then, the estimated regression line is $\hat{y} = 24.8544 + 4.4922x$. Note that the logarithm used here is the natural logarithm.

(b) The least squares estimates of the expected chemical influx at these values are

    i. $\hat{y}_{100} = 24.8544 + 4.4922\log(100) = 45.5417$ and

    ii. $\hat{y}_{100} = 24.8544 + 4.4922\log(150) = 47.3631$.

(c) By definition the residual sum of squares based on the *least squares* prediction line is the smallest of the residual sum of squares based on any other line. The visually estimated line is not a least squares line and therefore will have a residual sum of squares that is at least as large as the one based on the least squares prediction line. The $SS[E]$ based on the least squares line is

$$SS[E] = S_{yy} - \hat{\beta}_1^2 S_{xx} = 2205.7436 - (4.4922)^2(106.1659) = 63.3305.$$

The prediction equation from Example 10.2.2 is $\hat{y} = 24 + 5\log(\text{dose})$. The predicted value for the first observation is $24 + 5\log(1) = 24$. The residual is then $(y_{11} - \hat{y}_{11}) = 21 - 24 = -3$. The entire set of residuals for this equation is

−3.0000	0.0000	2.0000	1.0000	0.4871	2.4871	0.4871
−0.5129	−4.0258	−0.0258	−2.0258	1.9741	−4.5388	−2.5388
−0.5388	−5.5388					

The sum of squares for these residuals is 103.1468 which is larger than 63.30.

(d) The regression analysis used assumes a linear relationship between the expected influx and log(dose) while the ANOVA analysis assumes a relationship that includes straight line and other relationships. However, if the linearity assumption is not valid, then the $SS[E]$ will not be as small as it could be if the correct relationship was assumed. The ANOVA analysis does not have this disadvantage since it permits nonlinear relationships.

The predicted values for influx in the ANOVA analysis are simply the treatment means

Dose	1	10	100	1000
Mean	24	36.25	46	55.25

Hence, the residual for the first observation is $(Y_{11} - \overline{Y}_1) = 21 - 24 = -3$. The complete set of residuals for the ANOVA model is

−3.0	0.0	2.0	1.0	−0.25	1.75	−0.25	−1.25
−3.0	1.0	−1.0	3.0	−1.25	0.75	2.75	−2.25

From these, the residual sum of squares is 53.50 which is smaller than 63.30.

10.11. (a) The summary statistics for the analysis of mother's weight gain as a linear function of mother's age (in months) are
$$\bar{x} = 235.5000$$
$$\bar{y} = 10.2688$$
$$S_{xx} = 11502.0008$$
$$S_{yy} = 112.2525$$
$$S_{xy} = 1038.2208.$$

From these, the expected change in mother's gain in weight per 1 month increase in age can be estimated as $\hat{\beta}_1 = S_{xy}/S_{xx} = 1038.2208/11502.0008 = 0.0903$.

(b) The expected gain in weight of a mother aged 18 years can be calculated with the following least squares line.
$$\hat{y} = -10.9904 + 0.0903x$$
where $\hat{y}$ is the predicted weight gain and x is the mother's age in months. Thus, the estimate of the expected gain in weight of a mother aged 18 years is the predicted value at $x = 18$, i.e. $\hat{y} = -10.9904 + 0.0903(18 * 12) = 8.5144$.

(c) The error variance can be estimated as
$$SS[E] = S_{yy} - \hat{\beta}_1^2 S_{xx}$$
$$= 112.25249 - (0.0903)^2(11502.0008)$$
$$= 18.4641$$
$$\hat{\sigma}^2 = MS[E] = \frac{SS[E]}{n-2}$$
$$= \frac{18.4641}{8-2} = 3.0774$$

This value measures the variability of the observed values for weight gain around the least squares predicted values.

10.13. (a) In Exercise **10.9** the summary statistics were calculated. The construction of the 95% confidence interval for the expected chemical influx when the drug dose is equal to 100 nM is based on the formula

$$\hat{\mu}(x_0) \pm t(n-2, \alpha/2)\sqrt{MS[E]\left(\frac{1}{n} + \frac{(x_0 - \bar{x})^2}{S_{xx}}\right)}$$

where $\hat{\mu}(x_0) = 24.8544 + 4.4922\log(100) = 45.5417$ and $t(14, 0.025) = 2.1448$. MS[E] can be obtained either from Exercise **10.9** or Exercise **10.17**. Note that these two estimates are different because of roundoff. The former value is used here. Then, the formula becomes

$$45.5417 \pm 2.1448\sqrt{4.5236\left(\frac{1}{16} + \frac{(4.6052 - 3.4550)^2}{106.1659}\right)}$$

This becomes $(44.2926, 46.7907)$ which, at the 95% confidence level, is a range of plausible values for the expected chemical influx when the drug dose is 100 nM.

(b) Let $\mu(d_i)$ be the mean chemical influx when dose $= d_i$ for $i = 1, 2$. Then the change in the mean chemical influx that results from changing the drug dose from d_2 to d_1 is

$$\theta = \mu(d_2) - \mu(d_1) = \beta_0 + \beta_1 \log d_2 - (\beta_0 + \beta_1 \log d_1)$$

$$= \beta_1 \log\left(\frac{d_2}{d_1}\right)$$

$$= 0(\beta_0) + \log\left(\frac{d_2}{d_1}\right)\beta_1.$$

When $d_2 = 4d_1$, θ is a linear function with $c_0 = 0$ and $c_1 = \log\left(\frac{d_2}{d_1}\right) = \log 4 = 1.3863$. Since $\hat{\beta}_1 = 4.495$, $\hat{\theta} = 4.495(1.3863) = 6.2314$. The variance of $\hat{\beta}_1$ is

$$\hat{\sigma}_{\hat{\theta}}^2 = \hat{\sigma}^2\left\{s_{11}c_1^2\right\} = 4.5214\left(\frac{1}{113.1075}\right)1.9218 = 0.0768.$$

The hypothesis test that would verify the claim is (using $\alpha = 0.05$)

H_0: $\theta = 5$

H_1: $\theta > 5$

Test Statistic: $t_c = \dfrac{\hat{\theta} - 5}{\hat{\sigma}_{\hat{\theta}}^2} = \dfrac{6.2314 - 5}{\sqrt{0.0768}} = 4.4428$

Rejection Region: Reject H_0 if $t_c > t(14, 0.05) = 1.761$.

Conclusion: At the 5% level of significance, there is sufficient evidence to conclude that the chemical influx will increase by more than 5 nmol/mg protein if the drug dose is quadrupled.

The confidence interval approach to verify the claim will require us to calculate a lower bound for the change in mean chemical influx.

The lower bound is

$$\hat{\theta} - 1.761\hat{\sigma}_{\hat{\theta}} = 6.2314 - 1.761(0.2771) = 5.7434.$$

Since the lower bound is greater than 5, $\theta_0 = 5$ is outside the 95% lower confidence interval.

(c) The 95% confidence interval would be

$$\overline{Y}_i \pm t(df, \alpha/2)\sqrt{\frac{MSE}{n_i}}$$

$$46 \pm 2.179\sqrt{\frac{4.4583}{4}}$$

This interval becomes $(43.6995, 48.3004)$. The interval in Part a in Exercise **10.13** based on the regression analysis was $(44.3697, 46.7303)$. Because of the extra assumption of linearity that was made about the relationship, the precision is higher and thus the interval is narrower for the regression analysis.

(d) The figure shows the 90% confidence bands for the expected chemical influx. The 90% confidence intervals for the expected chemical influxes for drug doses $d = 100$ and $d = 150$ are $(40.7270, 50.3724)$ and $(42.2274, 52.5171)$, respectively. At the 90% confidence level we believe that the expected chemical influx is between 40.7270 and 40.3724 nmol/mg protein when the drug dose is 100 nM. Also at the 90% confidence level, we believe that the expected chemical influx is between 42.2274 and 52.5171 nmol/mg protein when the drug dose is 150 nM. Since each interval is at the 90% confidence level, the Bonferroni simultaneous confidence level is $1 - \alpha^*$, where $\frac{\alpha^*}{k} = \frac{\alpha^*}{2} = 0.10 \Rightarrow \alpha^* = 0.20$. Thus, we have at least 80% confidence that the two intervals will contain their parameters.

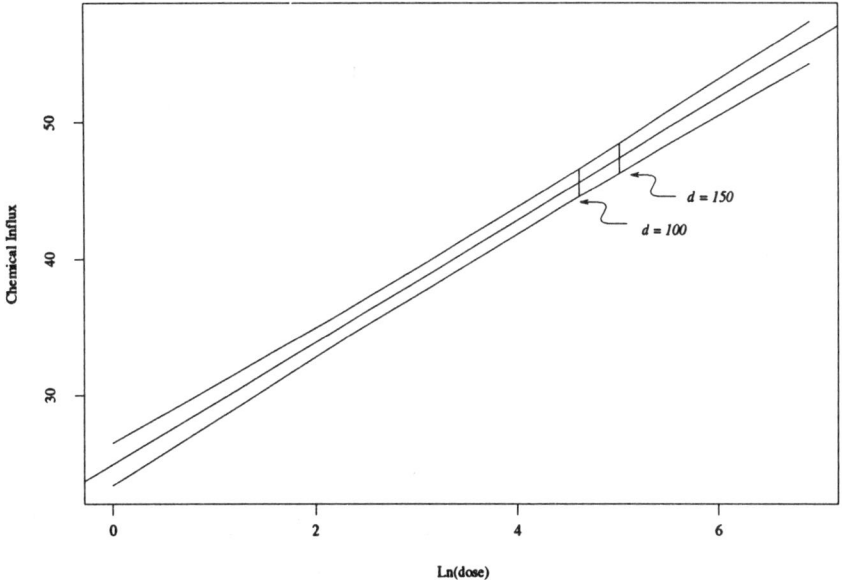

(e) The 99% lower bound for the mean chemical influx for the population of all rats receiving a drug dose equal to 10 nM is

$$24.8544 + 4.4922\log 10 - 2.624\sqrt{4.5236\left(\frac{1}{16} + \frac{(\log 10 - 3.4550)^2}{106.1659}\right)} = 33.6478.$$

This would be interpreted at the 99% confidence level as implying that the mean chemical influx for the population of all rats receiving a drug dose equal to 10 nM is at least as large as 33.6478.

(f) Given the simple linear regression model drug dose and chemical influx are associated if the slope β_1 is not equal to zero. The standard error of the slope is

$$\hat{\sigma}_{\hat{\beta}_1} = \hat{\sigma}\sqrt{\frac{1}{S_{xx}}} = 0.2065$$

95

The hypothesis test would be

H_0: $\beta_1 = 0$

H_1: $\beta_1 \neq 0$

Test Statistic: $t_c = \dfrac{4.4949}{0.2065} = 21.768$

Rejection Region: Reject H_0 if $t_c < -t(14, 0.025) = -2.145$ or $t > 2.145$.

Conclusion: At the 0.05 significance level, there is sufficient evidence to conclude that drug dose and chemical influx are associated.

(g) When the dose $d = 1$ $x = 0$. Then, the mean chemical influx for rats at $x = 0$ is the intercept. The question then asks us to test H_0: $\beta_0 \geq 25$ vs. H_1: $\beta_0 < 25$. The standard error for $\hat{\beta}_0$ is

$$\hat{\sigma}_{\hat{\beta}_0} = \hat{\sigma}\sqrt{\left(\frac{1}{n} + \frac{\bar{x}^2}{S_{xx}}\right)} = \sqrt{4.5236\left(\frac{1}{16} + \frac{3.4550^2}{106.1659}\right)} = 0.8896.$$

Then the hypothesis test is

H_0: $\beta_0 \geq 25$

H_1: $\beta_0 < 25$

Test Statistic: $t_c = \dfrac{\hat{\beta}_0 - 25}{\hat{\sigma}_{\hat{\beta}_0}} = \dfrac{24.8544 - 25}{0.8896} = -6.1097$

Rejection Region: Reject H_0 at the 0.05 significance level if $t_c < -t(14, 0.05) = 1.7613$

Conclusion: At the 0.05 significance level we reject H_0. The data does show that, on the average, rats receiving a dose $d = 1$ will have a chemical influx of less than 25 nM.

10.15. (a) The point estimate of the expected change in a mother's gain in weight for a 6 months increase in her age is $\hat{\theta} = 0(-11.2964) + 0.0922(6) = 0.5532$. Then, using the information in Box 10.5, the standard error is (with $c_0 = 0$ and $c_1 = 6$)

$$\hat{\sigma}_{\hat{\theta}} = \sqrt{\hat{\sigma}^2 \left\{ s_{00}c_0^2 + 2s_{01}c_0c_1 + s_{11}c_1^2 \right\}}$$

$$= \sqrt{4.0325\left\{ \left(\frac{1}{10367.9898}\right)6^2 \right\}}$$

$$= 0.1183$$

Then the 95% lower bound for the expected change in a mother's gain in weight for a 6 months increase in her age is $0.5532 - 1.943(0.1183) = 0.3233$. Hence, at the 95% confidence level, we may conclude that the expected change in a mother's gain in weight for a 6 months increase in her age is at least 0.3233 kg.

(b) Using the model with age in months, the point estimate for the expected gain in weight of a mother aged 35 years, equal to 420 months, is
$\hat{\mu}(420) = -11.2964 + 0.09216(420) = 27.41$. The standard error is

$$\hat{\sigma}_{\hat{\mu}(420)} = \sqrt{4.0325\left(\frac{1}{8} + \frac{(420 - 234.0)^2}{10367.9898}\right)} = 3.7363.$$

Therefore, the 95% confidence interval for the expected gain in weight of a mother aged 35 years is $27.41 \pm 2.447(3.7363)$ which is $(18.2673, 36.5527)$. At the 95% confidence level, 35 year old mothers should, on the average, gain between 18.2673 and 36.5527 kilograms.

Notice, however, that the range of the years that were actually observed is 15 to 23 years. 35 years is far outside the range of observed ages. The data have shown that a straight line models the relationship well between age and weight gain in the range 15 to 23 years. However, the data gives us no information about the relationship outside of this range. The relationship could continue in a straight line or it could curve in any direction. Our interval is *only* valid if the relationship continues as a straight line when age is greater than 23 years. If we have no evidence that this is true, then extrapolating to 35 years is dangerous.

(c) Since we are predicting the weight gain of a single 24 year old mother (288 months) the standard error is

$$\hat{\sigma}_{P(288)} = \sqrt{4.0325 \left(1 + \frac{1}{8} + \frac{(288 - 234.0)^2}{10367.9898}\right)} = 2.3813.$$

The point estimate is $\hat{\mu}(288) = -11.2963 + 0.09216(288) = 15.2458$ and $t(6, 0.01) = 3.143$. Then the upper limit is $15.2458 + 3.143(2.3813) = 22.7303$. At the 99% confidence level we can say that a 24 year old mother should gain at most 22.7303 kg in weight.

Note that 24 years is again outside the range of the data. However, it is much closer to the range than 35 years is. A researcher might feel more comfortable extrapolating to 24 years than 35 years.

10.17. (a) The following ANOVA tables were generated using PROC REG in SAS.

i. With $X = d$ the ANOVA table is

Dependent Variable: INFLUX

Analysis of Variance

Source	DF	Sum of Squares	Mean Square	F Value	Prob>F
Model	1	1348.73020	1348.73020	22.032	0.0003
Error	14	857.01980	61.21570		
C Total	15	2205.75000			

ii. With $X = \log(d)$ the ANOVA table is

Dependent Variable: INFLUX

Analysis of Variance

Source	DF	Sum of Squares	Mean Square	F Value	Prob>F
Model	1	2142.37651	2142.37651	473.278	0.0001
Error	14	63.37349	4.52668		
C Total	15	2205.75000			

(b) The coefficient of determination, R^2, can be calculated from the Sums of Squares given in the ANOVA table as

$$R^2 = \frac{SS[R]}{SS[TOT]}.$$

For the two ANOVA tables above, the coefficients of determination are

i. $R^2 = 1348.7302/2205.7500 = 0.6115$, and

ii. $R^2 = 2142.3765/2205.7500 = 0.9713$.

97

The second model fits the data better because 97.13% of the variation in chemical influx is accounted for by $X = \log(d)$, while only 61.15% of the variation in chemical influx is accounted for by $X = d$. That is, since R^2 is (much) larger for the second model than for the first model, the second model fits the data better.

(c) The predicted values for the first observation for the two models are

      i. $\hat{y} = 34.2857 + 0.0220(1) = 34.3076$, and

      ii. $\hat{y} = 24.8548 + 4.4921\log(1) = 24.8548$.

The entire set of observed and predicted values for the two models is

	Model (a), $X = d$		Model (b), $X = \log(d)$	
Obs	Observed	Predicted	Observed	Predicted
1	21.0000	34.3076	21.0000	24.8548
2	24.0000	34.3076	24.0000	24.8548
3	26.0000	34.3076	26.0000	24.8548
4	25.0000	34.3076	25.0000	24.8548
5	36.0000	34.5050	36.0000	35.1866
6	38.0000	34.5050	38.0000	35.1866
7	36.0000	34.5050	36.0000	35.1866
8	35.0000	34.5050	35.0000	35.1866
9	43.0000	36.4781	43.0000	45.5634
10	47.0000	36.4781	47.0000	45.5634
11	45.0000	36.4781	45.0000	45.5634
12	49.0000	36.4781	49.0000	45.5634
13	54.0000	56.2093	54.0000	55.8952
14	56.0000	56.2093	56.0000	55.8952
15	58.0000	56.2093	58.0000	55.8952
16	53.0000	56.2093	53.0000	55.8952
s^2	147.05	89.9153	147.05	142.8251

The variances of each column are given in the last row. These can be used to calculate the coefficient of determination via

$$R^2 = \frac{s^2_{pred}}{s^2_{obs}}.$$

The coefficients of determination calculated with this formula are exactly the same as the coefficients of determination calculated above. For example, the coefficient of determination for Model (a) is

$$R^2 = \frac{89.9153}{147.05} = 0.6115.$$

10.19. The summary statistics and parameter estimates for the weight gain data as calculated in Exercise 10.11 are

$$\bar{x} = 235.5000$$
$$\bar{y} = 10.2688$$
$$S_{xx} = 11502.0008$$
$$S_{yy} = 112.2525$$
$$S_{xy} = 1038.2208$$
$$\hat{\beta}_1 = 0.0903$$

To set up the ANOVA table realize that

$$SS[TOT] = S_{yy} = 112.2525,$$
$$SS[R] = \hat{\beta}_1^2 S_{xx} = (0.0903)^2(11502.0008) = 93.7884, \text{ and}$$
$$SS[E] = SS[TOT] - SS[R] = 112.2525 - 93.7884 = 18.4641.$$

The ANOVA table is

Source	DF	SS	MS	F
Regression	1	93.7884	93.7884	30.4770
Error	6	18.4641	3.0773	
Total	7	112.2525		

$$R^2 = \frac{SS[R]}{SS[TOT]} = \frac{93.7884}{112.2525} = 0.8355.$$

Thus, 83.55% of the variation in the observed weight gain can be accounted for by the straight line relationship between the expected weight gain and the mother's age in months. Note, that the value for R^2 would be exactly the same if the mother's age were used in years.

10.21. Referring back to the original data set from which this Example came, it is shown that:

$$\begin{aligned} n &= 10, \\ \bar{x} &= 207.0, \\ \bar{y} &= 50.3, \\ S_{xx} &= 80,266, \\ S_{yy} &= 3,288, \\ S_{xy} &= 15,283. \end{aligned}$$

It is also shown that the estimated regression line is:

$$\hat{y} = 10.89 + .19x \quad \text{and} \quad MS[E] = 47.29.$$

To compare and contrast the 90% prediction interval the following table will be very useful.

x_0	100	200
$\hat{\mu}(x_0)$	$10.89+.19(100)$ $=12.79$	$10.89+.19(200)$ $=14.69$
$\hat{\sigma}(x_0)$	$\sqrt{\left\{1+\frac{1}{10}+\frac{(100-207)^2}{80266}\right\}(47.29)}$ $=7.6658$	$\sqrt{\left\{1+\frac{1}{10}+\frac{(200-207)^2}{80266}\right\}(47.29)}$ $=7.2144$
One–at–a–Time	$12.79 \pm 1.86(7.6658)$ $(-1.4684, 27.0484)$	$14.69 \pm 1.86(7.2144)$ $(1.2712, 28.1088)$
Bonferroni	$12.79 \pm 2.306(7.6658)$ $(-4.8873, 30.4673)$	$14.69 \pm 2.306(7.2144)$ $(-1.9464, 31.3264)$
Working- Hotelling	$12.79 \pm 2.495(7.6658)$ $(-6.3372, 31.9162)$	$14.69 \pm 2.495(7.2144)$ $(-3.3099, 32.690)$

The One–at–a–Time method gives us at least 90% confidence that the average age of rats with body weights 100 g is in the interval -1.4684, 27.0484. A similar interpretation exists for rats with 200 g. These intervals do not tell us anything about what is going on simultaneously for the two means. The Bonferroni and Working-Hotelling intervals give 90% confidence that both the means are simultaneously within their respective intervals. Notice that these intervals are much wider than the confidence intervals for the One–at–a–Time expected response. Of the latter two, the Bonferroni intervals are better because they are narrower.

10.23. (a) In the altitude range of 0.64 to 1.72, the 95% Working-Hotelling band is

$$23.3543 - 8.1675x \pm 2.8636\sqrt{\left\{\frac{1}{12} + \frac{(x - 1.1300)^2}{1.6584}\right\} 7.04336} \quad 0.64 \le x \le 1.72.$$

The plot of this band is

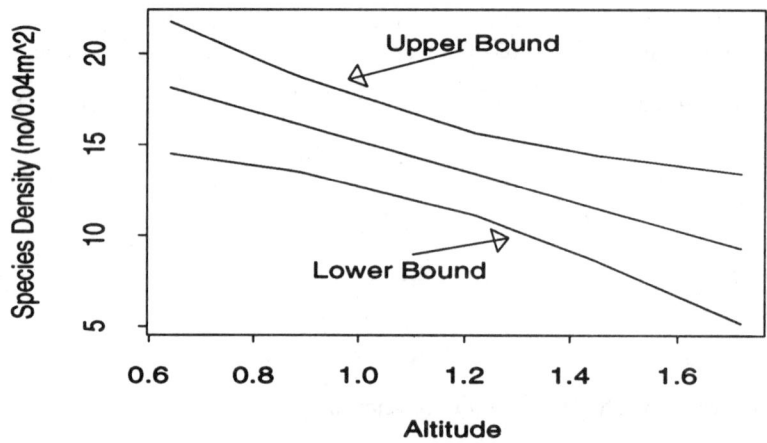

(b) The straight lines with the maximum and minimum slopes (in absolute value) can be found from the values of the confidence band endpoints at the altitudes 0.64 and 1.72. The straight line with the maximum slope will connect the upper bound at 0.64, 21.7568, and the lower bound at 1.72, 5.1908. The slope of the line connecting these two points is

$$\text{slope} = \frac{(5.1908 - 21.7568)}{(1.72 - 0.64)} = -15.3389.$$

The straight line with the minimum slope will connect the lower bound at 0.64, 14.4974, and the upper bound at 1.72, 13.4216. The slope of the line connecting these two points is

$$\text{slope} = \frac{(13.4216 - 14.4974)}{(1.72 - 0.64)} = -0.9961.$$

(c) At the 95% confidence level the slope of the line representing the relationship between species density and altitude is at least -15.3389 and at most -0.9961. In other words, at the 95% confidence level when altitude increases by 1000 meters the mean species density could decrease by as much as 15.3389 (no/0.04 mg^2) or by as little as 0.9961 (no/0.04 mg^2).

10.25. The following plots will help to check for assumption violations.

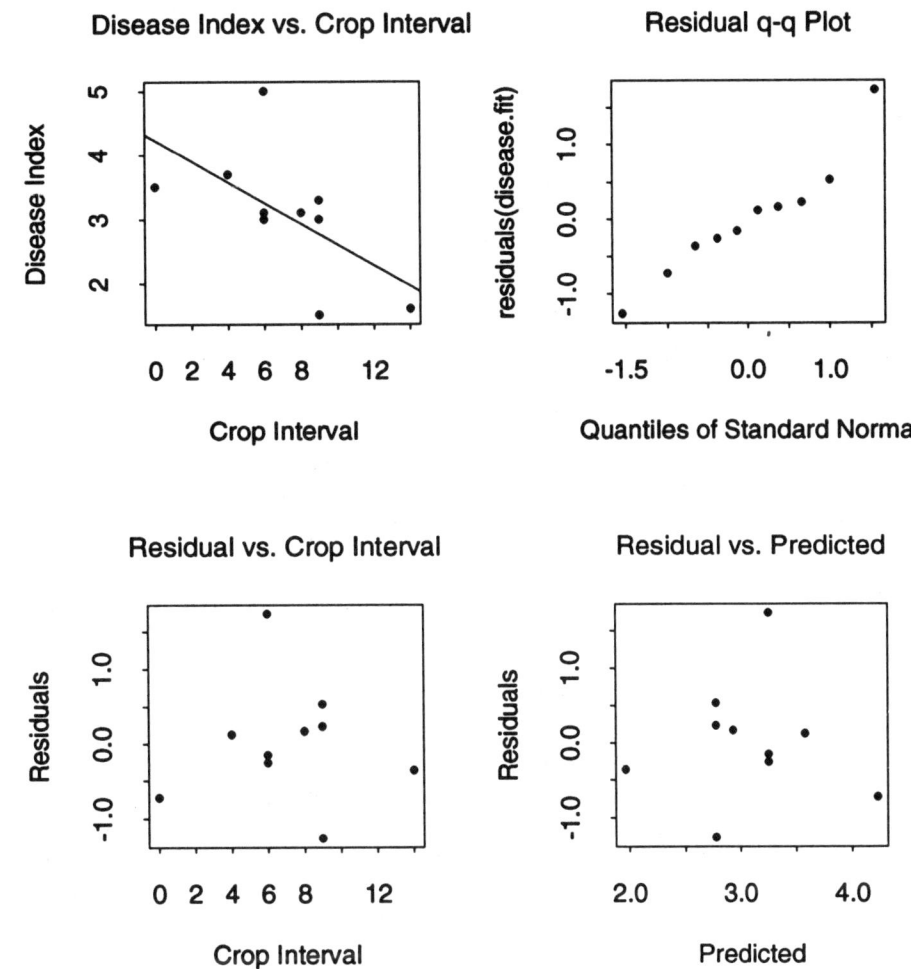

Based on the plots there do not seem to be any assumptions violated. Each of these plots is about what would be expected for a good fitting simple linear model.

10.27. The figure below shows the normal(q-q) plot, the residual vs. predicted plot and the residual vs. independent variable plot.

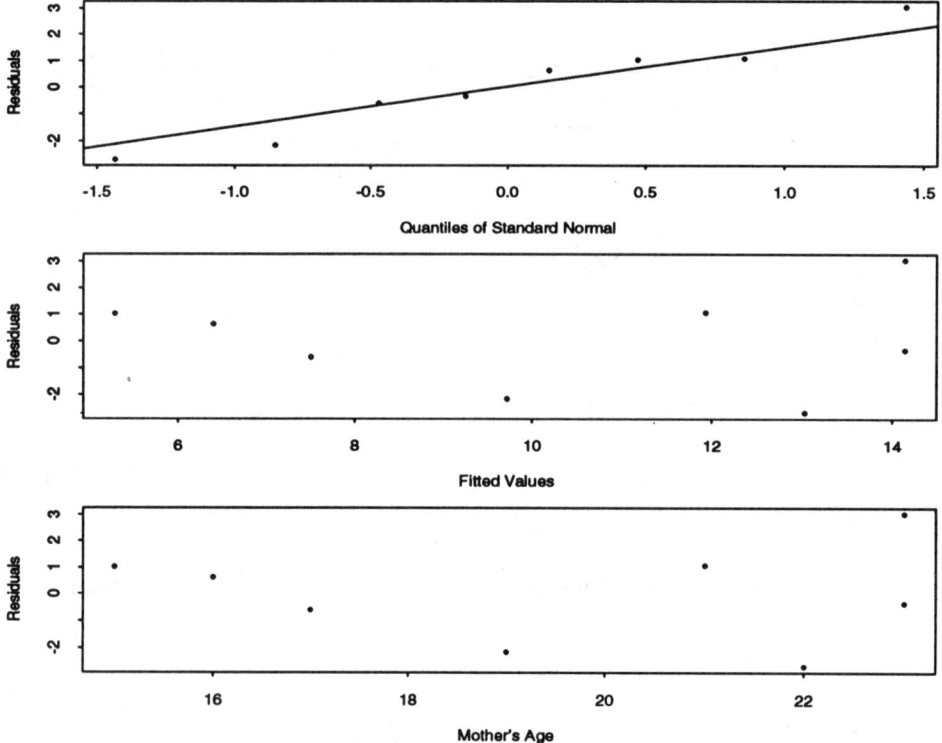

There does not appear to be any evidence of violations in the assumptions. The qqnorm does not show an exact straight line, but with the number of data points being used to fit the model, there is not any strong evidence of deviations from normality. The same is true for the other two plots. More data points would give a clearer picture of the distribution of the residuals. As it is, there is no compelling reason not to believe that the assumptions are being met.

10.29. (a) The simple linear regression lines are

      i. $\hat{\mu}(\ age) = -10.9904 + 1.0833\ age$, and

      ii. $\hat{\mu}(\ gain) = 11.7097 + 0.7708\ gain$.

(b) The correlation coefficient can be calculated in a number of ways. The estimated regression lines in Part 1 allow us to use

$$
\begin{aligned}
r_{XY} &= \pm\sqrt{\hat{\beta}_{Y.X}\hat{\beta}_{X.Y}} \\
&= \pm\sqrt{(1.0833)(0.7708)} = 0.9138.
\end{aligned}
$$

where the positive sign is used because both slopes are positive.

(c) The conversion from kilograms to pounds is $2.2046\,(kilograms) = pounds$. The conversion from years to months is $12\,(years) = months$. With these in mind, we have the following.

      i. The units for the slope $\hat{\beta}_{Y.X}$ in Part 1 are kg/yrs. The slope of the data that has been converted to pounds and months can be determined as

$$
\hat{\beta}_{new} = 1.0833\frac{\text{kg}}{\text{yr}} \times \frac{1yr}{12\text{months}} \times \frac{1\text{pound}}{2.2046\text{kg}} = 0.0409\frac{\text{pounds}}{\text{month}}.
$$

      ii. The correlation coefficient is unit-free. Thus converting the data from *kg* to *pounds* and *years* to *months* does not change the underlying relationship of the variables. Thus the correlation coefficient would still be 0.9138.

(d) The coefficient of determination can be obtained from the square of the correlation coefficient,

$$r_{XY}^2 = (r_{XY})^2 = (0.9138)^2 = 0.8350.$$

This tells us that 83.50% of the variation in the observed weight gain can be explained by a linear relationship between the expected weight gain and the mother's age.

Then, $1 - r_{XY}^2 = 1 - 0.8350 = 0.1650$ means that using mother's age as a predictor in a simple linear regression model reduced the variation of the observed weight gain by 16.50%.

(e) No linear association between the mother's age and her gain in weight means that $\rho_{XY} = 0$. Hence, we can test for the existence of a linear association with the following hypothesis test.

H$_0$: $\rho_{XY} = 0$

H$_1$: $\rho_{XY} \neq 0$

Test Statistic: $t_{r_{XY}} = \dfrac{r_{XY}\sqrt{n-2}}{\sqrt{1-r_{XY}^2}} = \dfrac{0.9138\sqrt{6}}{\sqrt{0.1650}} = 5.5104.$

Rejection Region: Reject H$_0$ in favor of $H_1 : \rho_{XY} \neq 0$ if
$$t_{r_{XY}} \leq -t(n-2, \alpha/2) = -t(6, 0.05) = -1.943 \text{ or } t_{r_{XY}} \geq 1.943$$

Conclusion: Since the test statistic falls in the rejection region, we reject H$_0$. At the 0.05 level of significance there is sufficient evidence to indicate that a linear association exists between a mother's age and her gain in weight during pregnancy.

10.31. The important part of the formula for the sample correlation coefficient,

$$r_{XY} = \frac{S_{xy}}{\sqrt{S_{xx}S_{yy}}},$$

is the numerator. $S_{xy} = \sum_{i=1}^{n} (x_i - \bar{x})(y_i - \bar{y})$. For both variables, the mean is 0. Thus, this equation reduces to $S_{xy} = \sum_{i=1}^{n} x_i y_i$. Because of the perfect symmetry in each variable each negative product, $x_i y_i$, has a positive counterpart (or vice versa) that "cancels" it out. For example, $(-3)(21) = -63$ cancels with $(3)(21) = 63$. Hence,

$$
\begin{aligned}
S_{xy} &= (-3)(21) + (-2)(11) + (-1)(5) + (0)(0) + (1)(5) + (2)(11) + (-3)(21) \\
&= 0.
\end{aligned}
$$

Finally, $S_{xy} = 0$ means that $r_{XY} = 0$.

10.33. Using Fisher's transformation we have,

$$
\begin{aligned}
Z_{XY} &= \frac{1}{2} \log \left(\frac{1 + r_{XY}}{1 - r_{XY}} \right) \\
&= \frac{1}{2} \log \left(\frac{1 + 0.393}{1 - 0.393} \right) \\
&= 0.4153
\end{aligned}
$$

(a) To answer the question we want to test H$_0$: $\rho_{XY}^2 \leq 0.30$ vs. H$_1$: $\rho_{XY}^2 > 0.30$. Hence, a lower confidence interval for ρ_{XY} is needed. This is of the form

$$Z_{XY} - z(\alpha) \frac{1}{\sqrt{n-3}}.$$

For $\alpha = 0.02$, $z(0.02)$ is approximately 2.05. The lower bound is

$$0.4153 - 2.05(0.1170) = 0.1754.$$

Using the inverse Fisher transformation,

$$
\begin{aligned}
r_{XY} &= \frac{e^{2Z_{XY}} - 1}{e^{2Z_{XY}} + 1} \\
&= \frac{e^{2(0.1754)} - 1}{e^{2(0.1754)} + 1} \\
&= 0.1736.
\end{aligned}
$$

Thus lower 98% confidence interval for the population correlation coefficient is $(0.1736, 1.0)$, where the upper bound is taken to be the maximum possible value for the parameter.

(b) The questions asks if there is support that the coefficient of determination is greater than 30%, i.e. $\rho^2 > 30$. The confidence interval for ρ^2 is $(0.1736, 1)^2 = (0.03, 1)$. THIS INTERVAL CONTAINS 0.30, so there is not support, at the 0.02 significance level, for the claim that more than 30% of the variation in the systolic blood pressures in the population can be explained by its linear association with the weights of the subjects in the population.

(c) Using a procedure similar to that in Part Part **a** in Exercise **10.33**, we calculate the 95% confidence interval for the correlation coefficient by first calculating a 95% confidence interval for ψ_{XY}.

$$
\left(Z_{XY} - z(\alpha/2)\frac{1}{\sqrt{n-3}}, Z_{XY} + z(\alpha/2)\frac{1}{\sqrt{n-3}} \right)
$$

With $z(0.025) = 1.96$ the interval is

$$
(0.4153 - 1.96(0.1170), 0.4153 + 1.96(0.1170)) = (0.1860, 0.6446).
$$

The 95% confidence interval for ρ_{XY} is obtained by applying the inverse Fisher transformation to the endpoints. That is,

$$
\frac{e^{2(0.1860)} - 1}{e^{2(0.1860)} + 1} \leq \rho_{XY} \leq \frac{e^{2(0.6446)} - 1}{e^{2(0.6446)} + 1}
$$
$$
\Rightarrow \quad 0.0186 \quad \leq \rho_{XY} \quad 0.1263
$$

This can be converted to a 95% confidence interval for the coefficient of determination by squaring each bound to get $(0.0003, 0.0159)$. Hence, with 95% confidence plausible values for the coefficient of determination for the prediction equation are between 0.03% and 1.59%.

10.35. (a) Box 10.10 defines $t_{r_{XY}}$ as

$$
t_{r_{XY}} = \frac{r_{XY}\sqrt{n-2}}{\sqrt{1 - r_{XY}^2}}
$$

The square of this is

$$
t_{r_{XY}}^2 = \frac{r_{XY}^2(n-2)}{1 - r_{XY}^2} = \frac{r_{XY}^2}{(1 - r_{XY}^2)/(n-2)}
$$

Now,

$$
r_{XY}^2 = \frac{SS_{Y.X}[R]}{SS_{Y.X}[TOT]}
$$

and

$$
1 - r_{XY}^2 = 1 - \frac{SS_{Y.X}[R]}{SS_{Y.X}[TOT]} = \frac{SS_{Y.X}[E]}{SS_{Y.X}[TOT]}
$$

Then,

$$t^2_{r_{XY}} = \frac{r^2_{XY}}{(1 - r^2_{XY})/(n - 2)} = \frac{\frac{SS_{Y.X}[R]}{SS_{Y.X}[TOT]}}{\left(\frac{SS_{Y.X}[E]}{SS_{Y.X}[TOT]}\right)/(n - 2)}$$

$$= \frac{SS_{Y.X}[R]}{SS_{Y.X}[E]/(n - 2)} = \frac{MS[R]}{MS[E]},$$

since there are 1 regression degrees of freedom and $n - 2$ error degrees of freedom. This relationship shows that test statistic for H_0: $\beta_1 = 0$, which follows the F-distribution with $(1, n - 2)$ degrees of freedom, if the null hypothesis is true, is the square of the test statistic for H_0: $\rho_{XY} = 0$, which follows the t-distribution with $(n - 2)$ degrees of freedom, if null hypothesis is true. It was shown in the text that a 2-sided t-test is equivalent to a one-sided F-test. This and the relationship above indicate that H_0: $\rho_{XY} = 0$ will be rejected if and only if H_0: $\beta_1 = 0$ is rejected. Hence, the two tests are equivalent.

(b) Using

$$t_{r_{XY}} = \frac{r_{XY}\sqrt{n - 2}}{\sqrt{1 - r^2_{XY}}}.$$

we have the following results.

i. For the first combination,

$$t_{r_{XY}} = \frac{0.7\sqrt{2 - 2}}{\sqrt{1 - (0.7)^2}}$$

which equals 0 because of the small sample size. The rest of the values are calculated in a similar way and are, with $r_{XY} = 0.7$,

Sample Size	2	4	6	10	50
$t_{r_{XY}}$	0	1.386207	1.960392	2.772413	6.790998

ii. With $n = 60$, and the following correlations $t_{r_{XY}}$ is

r_{XY}	0.2	0.4	0.6	0.8	0.9
$t_{r_{XY}}$	1.554563	3.323796	5.71183	10.154364	15.724604

The 0.05 critical values for (a) are

Sample Size	2	4	6	10	50
Critical Value	nonexistent	4.303	2.776	2.306	approx. 2.021

Comparing the test statistics in (a) with these critical values, the null hypothesis would not be rejected until the sample size is at least 10. This shows that a small sample size does not provide enough information for even a large correlation coefficient to be statistically significant.

The 0.05 critical value for (b) is 2.776. There is only one since there is only one sample size. Comparing 2.776 to the values of the test statistic in (b), the null hypothesis would be rejected when the correlation coefficient is only 0.4. Hence, a small observed correlation coefficient can be statistically significant if the sample size is large.

10.37. (a) The required sample size can be determined using the confidence interval approach with $t = 5, a = 150, b = 200, c_0 = 1, c_1 = 175, \alpha = .10, \sigma = 0.5$ or 0.6 in Equation (10.19). A SAS program and the resulting output is shown below.

```
--------------------------------------------------------------------
INPUT
data a;
```

```
input a b c0 c1  sigma t n @@;
lines;
150 200 1 175   .5  5 8 150 200 1 175   .5   5 7
150 200 1 175   .5  5 6 150 200 1 175   .5   5 5
150 200 1 175   .5  5 4 150 200 1 175   .5   5 3
150 200 1 175   .5  5 2 150 200 1 175   .5   5 1
150 200 1 175   .6  5 8 150 200 1 175   .6   5 7
150 200 1 175   .6  5 6 150 200 1 175   .6   5 5
150 200 1 175   .6  5 4 150 200 1 175   .6   5 3
150 200 1 175   .6  5 2 150 200 1 175   .6   5 1
;
data b;
  set a;
  R  = b-a;
  nu = n*t -2;
  x  = tinv(.95, nu );
  B  = x*(((t+1)*(c0)**2*(R**2) + 12*(t-1)*(c1 -
         0.5*c0*(a+b))**2)/(n*t*(t+1)*(R**2)))**0.5*(sigma);
put x;
put B;
proc print;
 var n t nu sigma x B;
run;
```

OUTPUT

OBS	N	T	NU	SIGMA	X	B
1	8	5	38	0.5	1.68595	0.13329
2	7	5	33	0.5	1.69236	0.14303
3	6	5	28	0.5	1.70113	0.15529
4	5	5	23	0.5	1.71387	0.17139
5	4	5	18	0.5	1.73406	0.19387
6	3	5	13	0.5	1.77093	0.22863
7	2	5	8	0.5	1.85955	0.29402
8	1	5	3	0.5	2.35336	0.52623
9	8	5	38	0.6	1.68595	0.15994
10	7	5	33	0.6	1.69236	0.17164
11	6	5	28	0.6	1.70113	0.18635
12	5	5	23	0.6	1.71387	0.20566
13	4	5	18	0.6	1.73406	0.23265
14	3	5	13	0.6	1.77093	0.27435
15	2	5	8	0.6	1.85955	0.35282
16	1	5	3	0.6	2.35336	0.63147

Thus, we will need between 2 (corresponding to $\sigma = 0.5$) and 3 replications (corresponding to $\sigma = 0.6$) at each design point.

(b) For this simultaneous estimation we can use the Bonferroni approach with $k = 2$. Proceed as in (a) to determine the sample sizes corresponding to $c_1 = 160, \alpha = 0.05$ and $c_1 = 170, \alpha = 0.05$. The maximum of these two sample sizes is the required sample size. Running the above SAS program under these conditions gives the (summarized) results

106

σ	c_1	n
0.5	160	3
0.6	160	4
0.5	170	2
0.6	170	3

Using the largest of these gives 4 replications for each design point.

(c) To determine the replication sizes for predicting the FEV of a single subject we can proceed as in (a) by replacing W in Equation (10.19) with

$$W^* = 2t(nt - 2, \alpha/2) \left[1 + \frac{(t+1)c_0^2 R^2 + 12(t-1)(c_1 - m_0 c_0)^2}{nt(t+1)R^2}\right]^{.5} \sigma.$$

However, these conditions do not provide a solution. As the sample size gets larger the above equality in the limit goes to

$$W^* = 2z(\alpha/2)[1]^{.5}\sigma.$$

However, the right hand side is 0.8225 and does not depend on n. The researcher cannot expect to achieve a bound on the error of estimation that is as small as 0.3 liters.

(d) Realizing the lower bound on the error of estimation that was found in Part Part c in Exercise **10.37** and supposing that the researcher was comfortable with a larger error of estimation, the we could proceed as in Part Part b in Exercise **10.37** by replacing W in Equation (10.19) with W^* defined in Part c in Exercise **10.37**.

10.39. The response is replicated at each level of dosage. This allows a measure of the variability of the response at each level of the independent variability. A plot of the chemical influxes versus the dosage will provide visual information about this variability (cf. Exercise 10.24). If a model is proposed, then the residuals can be plotted versus the fitted values. The information should be the same.

If the variances appear to be unequal, then reasonable weights would be to use the reciprocal of the sample variances, $1/s_i^2$.

10.41. For the disease index data, not all levels of Crop Index have multiple responses. Because of this, estimates of sample variances can not be found. Hence, no obvious way of estimating the weights can be found.

10.43. (a) For the pedagogical reasons *only* (in practice one would refer to Sen (1968)), we replace the observation $(180, 47)$ with $(179.5, 47)$.

Following the same calculations as in Exercise **10.42**, for example, we arrive at the ordered values for $\hat{\beta}_{ij}$ as

−0.8889	−0.8148	−0.7778	−0.6000	−0.3636
−0.3214	−0.1667	−0.0526	−0.0294	0.0000
0.1556	0.1707	0.2250	0.3429	0.4167
0.4688	0.5385	0.5833	0.8085	0.8125
0.8769	0.9412	0.9437	1.0361	1.0606
1.0886	1.0926	1.2105	1.2308	1.3765
1.5106	1.5833	1.6111	1.6667	1.8710
2.0714	2.0789	2.3750	2.4737	3.0000
3.0000	3.5714	5.1667	13.0000	15.5000

With $N = n(n-1)/2 = 45$ and from the Table C.13, $S(10, .025) \approx 23$

$$L = \frac{N - S(n, \frac{\alpha}{2})}{2} + 1$$

$$= \frac{45 - 23}{2} + 1 = 12$$

$$U = \frac{N + S(n, \frac{\alpha}{2})}{2}$$

$$= \frac{45 + 23}{2} = 34.$$

Then, the approximate 95% confidence interval for the slope parameter in the regression of a rat's age on its weight is $\left(\hat{\beta}_{(12)}, \hat{\beta}_{(34)} \right) = (0.0962, 0.2486)$.

(b) It is assumed that the rat's age is a linear function of its weight and random error. It is further assumed that the errors are independent with a common continuous distribution with median 0. Another way to say this is that the rat's ages are independent with a common continuous distribution with median equal to a linear function of the rat's weight.

(c) The test to be performed is an upper tailed test.

H$_0$: $\beta_1 = 0$
H$_1$: $\beta_1 > 0$
Test Statistic: $S_c = \left\{ \left(\text{Number of } \hat{\beta}_{ij} > 0 \right) - \left(\text{Number of } \hat{\beta}_{ij} < 0 \right) \right\} = 41 - 4 = 37$
Rejection Region: Reject H$_0$ if $S_c > S(10, .01) \approx 27$
Conclusion: Reject H$_0$. There is sufficient evidence at the 0.01 significance level to say that a positive association exists between the age and weight of rats.

(d) Using the results in Examples 10.3.2 and 10.3.3, we have the following statistics

$$\hat{\beta}_1 = 0.1904 \quad \text{and} \quad \hat{\sigma}_{\hat{\beta}_1} = \frac{\hat{\sigma}}{\sqrt{S_{xx}}} = \frac{47.29}{80266} = 0.0006.$$

The test is, then,

H$_0$: $\beta_1 = 0$
H$_1$: $\beta_1 > 0$
Test Statistic: $t_{\hat{\beta}_1} = \hat{\beta}_1 / \hat{\sigma}_{\hat{\beta}_1} = 0.1904/0.0006 = 323$. Comparing this to the $t(8)$ distribution the p-value ≈ 0.
Conclusion: Reject H$_0$. With a p-value ≈ 0, the data does show that a positive relationship exists.

The results of this test are consistent with the results of the rank regression test. We would want to use the rank regression when the assumptions for the simple linear regression method are not met.

10.45. Let $\beta_0 = 1.0$ and $\beta_1 = -1.2$. Then, for any value of $x > 0.833$, $P(x)$ will be negative.

10.47. (a) Assuming that the probability that a patient of age $X = x$ will experience pulmonary toxicity can be expressed in terms of two parameters, β_0 and β_1 a logistic regression model for relating the probability of pulmonary toxicity to the age of the patient would be

$$P(x) = \frac{e^{\beta_0 + \beta_1 x}}{1 + e^{\beta_0 + \beta_1 x}}.$$

(b) The interpretation of $\beta_0 = -1.6033$ is that the probability of pulmonary toxicity in a patient of "age 0" is estimated to be

$$
\begin{aligned}
\hat{P}(0) &= \frac{e^{-1.6033}}{1 + e^{-1.6033}} \\
&= 0.1675
\end{aligned}
$$

Because the estimated value of 0.0901 for β_1 is positive the probability of pulmonary toxicity in a patient increases as the age of the patient increases.

(c) The estimated probability of PT as a function of age is

$$
\hat{P}(x) = \frac{e^{-1.6033+0.0901x}}{1 + e^{-1.6033+0.0901x}}.
$$

(d) The estimated probability probability of PT for a patient aged 15 years is

$$
\begin{aligned}
\hat{P}(x) &= \frac{e^{-1.6033+0.0901(15)}}{1 + e^{-1.6033+0.0901(15)}} \\
&= 0.4374.
\end{aligned}
$$

Thus, a 15 year old patient has a 43.74% chance of having pulmonary toxicity.

(e) $p-$value $= 0.0008$ for β_0 and $p-$value $= 0.0226$ for β_1 imply that each of these coefficients is significantly different from zero.

(f) The 95% confidence interval for the probability of PT for a patient aged 13 years is given in the output as $(0.1906, 0.4104)$. With 95% confidence we can claim that between 19% and 41% of the 13 year old patients will have acute pulmonary toxicity with 100 days of BMT.

Chapter 11

Multiple Linear Regression

11.1. (a) Six plots of the cubic regression function at various values of the four parameters are shown below.

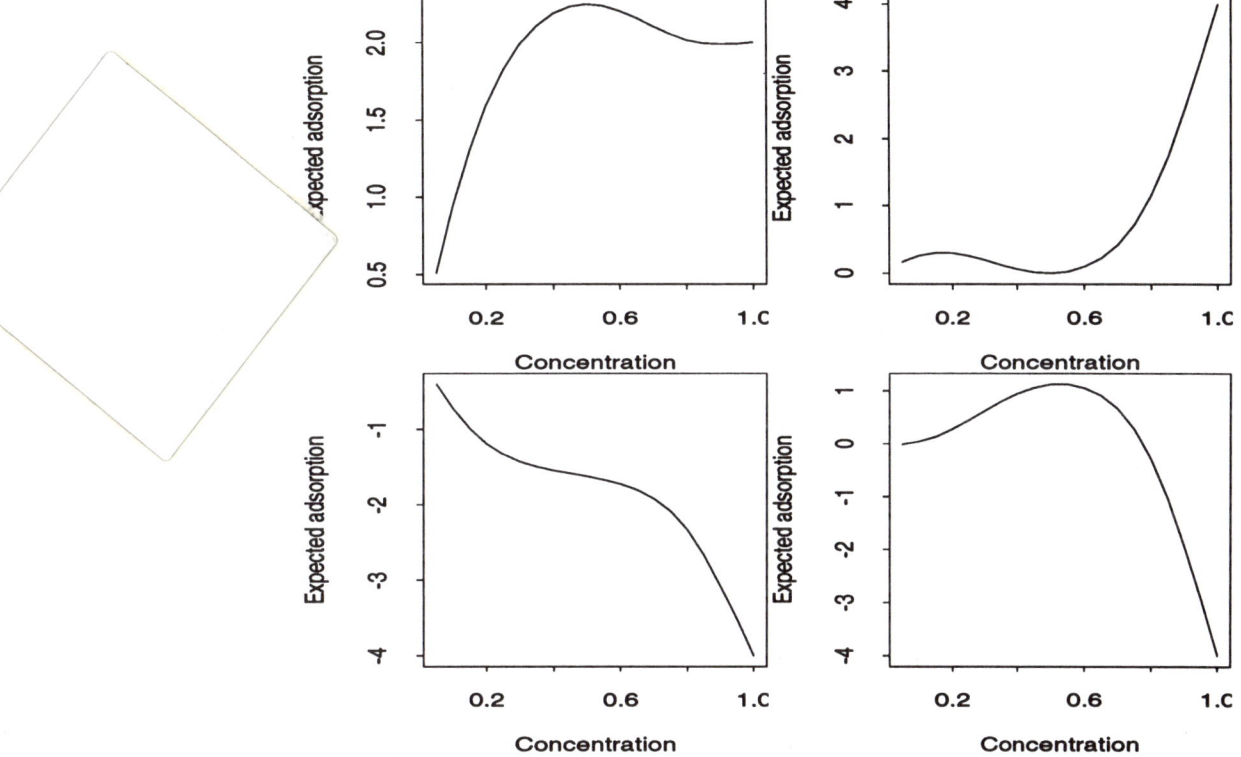

The cubic regression function has some features that make it less appropriate than the linear and quadratic regression functions as models of the relationship between the expected adsorption and the enzyme concentration. The quadratic and linear regression functions are more parsimonious than the cubic regression function, meaning that they are simpler. As can be seen in the above figure, the cubic plot generally has two bends in the line. It would be difficult to biologically explain these rises and dips.

(b) The cubic regression function in Part (1) can be represented as a multiple linear regression function with $p = 3$ independent variables in the following manner. Let

$$\mu(x_1, x_2, x_3) = E(Y|C = c) = \mu(c)$$
$$x_1 = c$$
$$x_2 = c^2$$
$$x_3 = c^3$$

The resulting multiple linear regression function is

$$\mu(x_1, x_2, x_3) = \beta_0 + \beta_1 x_1 + \beta_2 x_2 + \beta_3 x_3$$

11.3. (a) i. Additive, since the function can be expressed as the sum of five terms each containing only one explanatory variable.
   ii. Interactive, since the terms $\beta_4 ab$, $\beta_5 ac$, and $\beta_6 bc$ cannot be expressed as a sum of components that depend upon the level of at most a single factor.
   iii. Interactive. This function is the previous function with additive terms incorporated. New terms do not remove interaction.
   iv. Interactive, since many of the terms, for example $\beta_6 bc$, can not be expressed as a sum of components that depend upon the level of at most a single factor.

(b) β_1 is the change in the expected value of log(oxygen demand) as the biological oxygen demand, (A), increases by 1 unit, holding all the others constant.

β_2 is the change in the expected value of log(oxygen demand) as the total Kjeldahl nitrogen, (B), increases by 1 unit, holding all the others constant.

(c) $\beta_1 + \beta_2 + \beta_3$ is the change in the expected value of log(oxygen demand) as each of biological oxygen demand, (A), total Kjeldahl nitrogen, (B), and total solids, (C), increases by 1 unit each, holding total volatile solids, (D) and (F), constant.

(d) Using the notation in the text, the effect of increasing A alone by 1 unit is

$$\mu(\Delta A) = \mu(a + 1, b, c) - \mu(a, b, c) = \beta_1 + \beta_4 b + \beta_5 c.$$

The effect of increasing B alone by 1 unit is

$$\mu(\Delta B) = \mu(a, b + 1, c, d, f) - \mu(a, b, c, d, f) = \beta_2 + \beta_4 a + \beta_6 c.$$

And then, the effect of simultaneously increasing the values of both A and B by 1 unit each is

$$\mu(\Delta A, \Delta B) = \mu(a + 1, b + 1, c, d, f) - \mu(a, b, c, d, f)$$
$$= \beta_1 + \beta_2 + \beta_4 a + \beta_4 b + \beta_4 + \beta_5 c + \beta_6 c$$
$$= (\beta_1 + \beta_4 b + \beta_5 c) + (\beta_2 + \beta_4 a + \beta_6 c) + \beta_4$$
$$= \mu(\Delta A) + \mu(\Delta B) + \beta_4$$

Hence, β_4 is the difference between the effect of simultaneously increasing the values of both A and B and the sum of the effects of increasing them individually. The fact that these two effects are not equal is the definition of interaction and, thus, β_4 is a measure of the effect of the interaction between A and B.

11.5. In Exercise **11.2**, the four regression functions given were expressed as multiple linear regression functions. The multiple linear regression *models* that correspond to these functions are statistical models with the actual responses written in terms of the expected responses and the random error. The expected responses are represented by the multiple linear regression function given in Exercise **11.2**. These would be formally written as the following, with E_i, the random error in Y_i, assumed to be independent $N(0, \sigma^2)$,

(a) $Y_i = \beta_0 + \beta_1 x_{i1} + \beta_2 x_{i2} + \beta_3 x_{i3} + \beta_4 x_{i4} + \beta_5 x_{i5} + E_i, \qquad i = 1, \ldots, n.$

(b) $Y_i = \beta_0 + \beta_1 x_{i1} + \beta_2 x_{i2} + \beta_3 x_{i3} + \beta_4 x_{i5} + \beta_5 x_{i6} + \beta_6 x_{i7} + E_i, \qquad i = 1, \ldots, n.$

(c) $Y_i = \beta_0 + \beta_1 x_{i1} + \beta_2 x_{i2} + \beta_3 x_{i3} + \beta_4 x_{i4} + \beta_5 x_{i5} + \beta_6 x_{i6} + \beta_7 x_{i7} + \beta_8 x_{i8} + \beta_9 x_{i9} + E_i, \qquad i = 1, \ldots, n.$

(d) $Y_i = \beta_0 + \beta_1 x_{i1} + \beta_2 x_{i2} + \beta_3 x_{i3} + \beta_4 x_{i4} + \beta_5 x_{i5} + \beta_6 x_{i6} + \beta_7 x_{i7} + \beta_8 x_{i8} + \beta_9 x_{i9} + \beta_{10} x_{i10} + \beta_{11} x_{i11} + \beta_{12} x_{i12} + \beta_{13} x_{i13} + \beta_{14} x_{i14} + \beta_{15} x_{i15} + E_i, \qquad i = 1, \ldots, n.$

11.7. (a) $\boldsymbol{E} + \boldsymbol{O} = \boldsymbol{E}$ would be written as

$$\begin{bmatrix} 3 & 4 & 1 & 0 \\ 2 & 0 & 6 & 2 \end{bmatrix} + \begin{bmatrix} 0 & 0 & 0 & 0 \\ 0 & 0 & 0 & 0 \end{bmatrix} = \begin{bmatrix} 3+0 & 4+0 & 1+0 & 0+0 \\ 2+0 & 0+0 & 6+0 & 2+0 \end{bmatrix}$$

where the last matrix clearly equals $\boldsymbol{E}$.

(b) $\boldsymbol{OE'} = \boldsymbol{O}$ would be written as

$$\begin{bmatrix} 0 & 0 & 0 & 0 \\ 0 & 0 & 0 & 0 \end{bmatrix} \begin{bmatrix} 3 & 2 \\ 4 & 0 \\ 1 & 6 \\ 0 & 2 \end{bmatrix}$$

$$= \begin{bmatrix} (0)(3)+(0)(4)+(0)(1)+(0)(0) & (0)(2)+(0)(0)+(0)(6)+(0)(2) \\ (0)(3)+(0)(4)+(0)(1)+(0)(0) & (0)(2)+(0)(0)+(0)(6)+(0)(2) \end{bmatrix}$$

$$= \begin{bmatrix} 0 & 0 \\ 0 & 0 \end{bmatrix}$$

and $\boldsymbol{O'E} = \boldsymbol{O'}$ is

$$\begin{bmatrix} 0 & 0 \\ 0 & 0 \\ 0 & 0 \\ 0 & 0 \end{bmatrix} \begin{bmatrix} 3 & 4 & 1 & 0 \\ 2 & 0 & 6 & 2 \end{bmatrix}$$

$$= \begin{bmatrix} (0)(3)+(0)(4) & (0)(4)+(0)(0) & (0)(1)+(0)(6) & (0)(0)+(0)(2) \\ (0)(3)+(0)(4) & (0)(4)+(0)(0) & (0)(1)+(0)(6) & (0)(0)+(0)(2) \\ (0)(3)+(0)(4) & (0)(4)+(0)(0) & (0)(1)+(0)(6) & (0)(0)+(0)(2) \\ (0)(3)+(0)(4) & (0)(4)+(0)(0) & (0)(1)+(0)(6) & (0)(0)+(0)(2) \end{bmatrix}$$

$$= \begin{bmatrix} 0 & 0 & 0 & 0 \\ 0 & 0 & 0 & 0 \\ 0 & 0 & 0 & 0 \\ 0 & 0 & 0 & 0 \end{bmatrix}$$

Notice that the two resultant null matrices have different dimensions than the original null matrix.

(c) $FI = F$ would be written as

$$\begin{bmatrix} 0 & 2 & 6 \\ 1 & 4 & 0 \\ 3 & 0 & 1 \end{bmatrix} \begin{bmatrix} 1 & 0 & 0 \\ 0 & 1 & 0 \\ 0 & 0 & 1 \end{bmatrix}$$

$$= \begin{bmatrix} (0)(1)+(2)(0)+(6)(0) & (0)(0)+(2)(1)+(6)(0) & (0)(0)+(2)(0)+(6)(1) \\ (1)(1)+(4)(0)+(0)(0) & (1)(0)+(4)(1)+(0)(0) & (1)(0)+(4)(0)+(0)(1) \\ (3)(1)+(0)(0)+(1)(0) & (3)(0)+(0)(1)+(1)(0) & (3)(0)+(0)(0)+(1)(1) \end{bmatrix}$$

$$= \begin{bmatrix} 0 & 2 & 6 \\ 1 & 4 & 0 \\ 3 & 0 & 1 \end{bmatrix}.$$

$IG = G$ would be written as

$$\begin{bmatrix} 1 & 0 & 0 \\ 0 & 1 & 0 \\ 0 & 0 & 1 \end{bmatrix} \begin{bmatrix} 2 & 6 & 3 \\ 6 & 4 & 1 \\ 3 & 1 & 7 \end{bmatrix}$$

$$= \begin{bmatrix} (1)(2)+(0)(6)+(0)(3) & (0)(2)+(1)(6)+(0)(3) & (0)(2)+(0)(6)+(1)(3) \\ (1)(6)+(0)(4)+(0)(1) & (0)(6)+(1)(4)+(0)(1) & (0)(6)+(0)(4)+(1)(1) \\ (1)(3)+(0)(1)+(0)(7) & (0)(3)+(1)(1)+(0)(7) & (0)(3)+(0)(1)+(1)(7) \end{bmatrix}$$

$$= \begin{bmatrix} 2 & 6 & 3 \\ 6 & 4 & 1 \\ 3 & 1 & 7 \end{bmatrix}.$$

Notice that the F and G matrices do not change.

11.9. For the design matrix in Example 11.8, the matrices $X\beta$ and $X'y$ are

$$X\beta = \begin{bmatrix} 1 & 4.2 & 6.9 \\ 1 & 3.6 & 2.3 \\ 1 & 4.7 & 7.6 \\ 1 & 5.2 & 5.8 \\ 1 & 4.6 & 8.1 \\ 1 & 5.0 & 8.6 \\ 1 & 5.4 & 1.7 \\ 1 & 7.0 & 6.9 \end{bmatrix} \begin{bmatrix} \beta_0 \\ \beta_1 \\ \beta_2 \end{bmatrix} = \begin{bmatrix} \beta_0 + 4.2\beta_1 + 6.9\beta_2 \\ \beta_0 + 3.6\beta_1 + 2.3\beta_2 \\ \beta_0 + 4.7\beta_1 + 7.6\beta_2 \\ \beta_0 + 5.2\beta_1 + 5.8\beta_2 \\ \beta_0 + 4.6\beta_1 + 8.1\beta_2 \\ \beta_0 + 5.0\beta_1 + 8.6\beta_2 \\ \beta_0 + 5.4\beta_1 + 1.7\beta_2 \\ \beta_0 + 7.0\beta_1 + 6.9\beta_2 \end{bmatrix}$$

$$X'y = \begin{bmatrix} 1 & 1 & 1 & 1 & 1 & 1 & 1 & 1 \\ 4.2 & 3.6 & 4.7 & 5.2 & 4.6 & 5.0 & 5.4 & 7.0 \\ 6.9 & 2.3 & 7.6 & 5.8 & 8.1 & 8.6 & 1.7 & 6.9 \end{bmatrix} \begin{bmatrix} 1.7 \\ 1.9 \\ 2.1 \\ 2.8 \\ 2.2 \\ 2.9 \\ 3.4 \\ 4.1 \end{bmatrix} = \begin{bmatrix} 21.10 \\ 110.09 \\ 125.13 \end{bmatrix}$$

11.11. (a) Clearly, since β_0 is the value of $E(Y_i)$ when both x_{i1} and x_{i2} are equal to zero, β_0 has the same meaning in both models. However, Model 1 is an additive model and Model 2

114

is an interactive model. Thus, in Model 1, β_1 (β_2) is the change in $E(Y_i)$ as x_{i1} (x_{i2}) increases by 1 unit holding x_{i2} (x_{i1}) constant at any value. On the other hand, for Model 2, when x_{i2} is held fixed at x_2, the change in $E(Y_i)$ is $\beta_1 + \beta_2 x_2$ instead of just β_1.

(b) From the **Parameter Estimates** in the printout for Model 1 the estimated least squares prediction equation is $\hat{y} = -0.8611 + 0.8000 x_1 - 0.0788 x_2$.

(c) The error sum of squares, 0.06367, is equal to $\sum_{i=1}^{n}(y_i - \hat{y}_i)^2$. $\hat{y}_i$ is the ith predicted value from the fitted model and $y_i - \hat{y}_i$ is the ith residual. Thus the error sum of squares is the sum of the squared residuals. This can be verified from the values of y_i and $\hat{y}_i$.

(d) **H$_0$:** $\beta_1 = \beta_2 = 0$

H$_1$: At least one of β_1, $\beta_2 \neq 0$

Test Statistic: p–value $= 0.0012$

Conclusion: Since 0.0012 is smaller than 0.01 we reject the null hypothesis at the 0.01 level of significance. There is sufficient evidence to conclude that either $\beta_1 \neq 0$ or $\beta_2 \neq 0$ or both.

(e) From the **Parameter Estimates** in the printout for Model 2 the estimated least squares prediction equation is $\hat{y} = -0.5977 + 0.7435 x_1 - 0.1261 x_2 + 0.0100 x_1 x_2$.

(f) The predicted values for Model 2 are obtained by substituting the values for the independent variables into the estimated prediction equation in Part Part **e** in Exercise **11.11**. The residuals are then obtained by subtracting the predicted values from the observed values. For example, for observation 1

$$\hat{y}_1 = -0.5977 + 0.7435(4.2) - 0.1261(6.9) + 0.01(4.2)(6.9) = 1.9442$$

The observed value is 1.7, so that the residual at observation 1 is
$(y_1 - \hat{y}_1) = 1.7 - 1.9442 = -0.2442$.

The complete set of residuals for Model 2 are

	Dep Var	Predict Value	Residual
1	1.7000	1.9442	-0.2442
2	1.9000	1.8716	0.0284
3	2.1000	2.2950	-0.1950
4	2.8000	2.8382	-0.0382
5	2.2000	2.1729	0.0271
6	2.9000	2.4646	0.4354
7	3.4000	3.2947	0.1053
8	4.1000	4.2189	-0.1189

(g) The two plots of residuals versus observed values are almost the same. The largest residual for both models is for $y = 2.9$. Thus both models seem to be of very similar adequacy. Given this, then, model 1 is the preferred model because it is simpler.

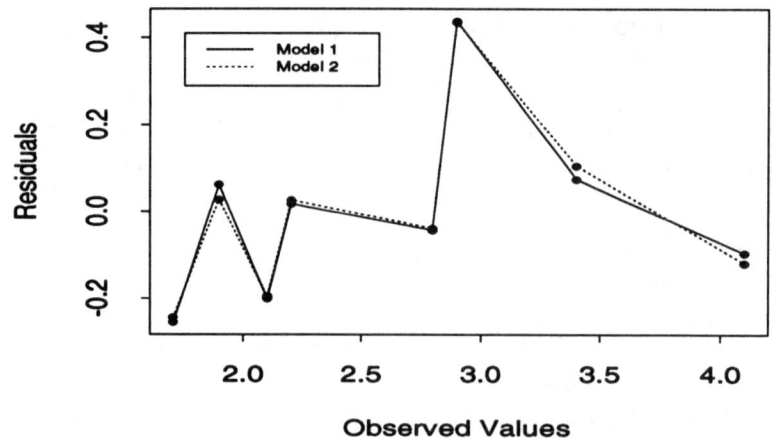

(h) The estimates of the error variances are $\hat{\sigma}^2 = MS[E] = 0.06367$ for Model 1 and $\hat{\sigma}^2 = MS[E] = 0.07887$ for Model 2. Which model should be preferred should be based on considerations such as the fit of the model and the underlying assumptions. Estimated error variance does not give useful information about the adequacy of a statistical model.

11.13. Looking at the general form of the normal equations a few things can be noticed. First, there are $p + 1$ equations corresponding to the $p + 1$ regression coefficients. In Example 11.4.1 $p = 2$ so there are 3 equations. There are also $p + 1$ terms in the sum on the right hand side of each equation; one for the intercept and one for each of the slopes. Again, since $p = 2$ there are 3 terms in the sum on the right hand side.

Each sum is taken over all the observations. There are 7 observations, $n = 7$, so each sum ranges over 1 to 7. Finally, in the general equations, notice that each independent variable is crossed with $y_k, 1$, and the other independent variables, including itself. This is the case in Example 11.9. The second equation has x_{i1} crossed with y_i on the left hand side. The first summand is x_{i1} times "1", the second summand is x_{i1} times itself, and the third summand is x_{i1} times x_{i2}. The other two equations are similar.

11.15. The requested matrices for the two models are

Model 1

X	$X'X$	$X'y$
$\begin{bmatrix} 1 & 4.2 & 6.9 \\ 1 & 3.6 & 2.3 \\ 1 & 4.7 & 7.6 \\ 1 & 5.2 & 5.8 \\ 1 & 4.6 & 8.1 \\ 1 & 5.0 & 8.6 \\ 1 & 5.4 & 1.7 \\ 1 & 7.0 & 6.9 \end{bmatrix}$	$\begin{bmatrix} 8.0 & 39.7 & 47.9 \\ 39.7 & 204.05 & 240.88 \\ 47.9 & 240.88 & 334.37 \end{bmatrix}$	$\begin{bmatrix} 21.1 \\ 110.09 \\ 125.13 \end{bmatrix}$

116

X	$X'X$	$X'y$
$\begin{bmatrix} 1 & 4.2 & 6.9 & 28.98 \\ 1 & 3.6 & 2.3 & 8.28 \\ 1 & 4.7 & 7.6 & 35.72 \\ 1 & 5.2 & 5.8 & 30.16 \\ 1 & 4.6 & 8.1 & 37.26 \\ 1 & 5.0 & 8.6 & 43.0 \\ 1 & 5.4 & 1.7 & 9.18 \\ 1 & 7.0 & 6.9 & 48.3 \end{bmatrix}$	$\begin{bmatrix} 8.0 & 39.7 & 47.9 & 240.88 \\ 39.7 & 204.05 & 240.88 & 1250.308 \\ 47.9 & 240.88 & 334.37 & 1685.888 \\ 240.88 & 1250.308 & 1685.888 & 8748.413 \end{bmatrix}$	$\begin{bmatrix} 21.1 \\ 110.09 \\ 125.13 \\ 660.372 \end{bmatrix}$

11.17. (a) Using the SAS printout in Exercise **11.11**, we have the following hypothesis test.

$\mathbf{H_0}$: $\beta_1 = 0$

$\mathbf{H_1}$: $\beta_1 \neq 0$

Test Statistic: p−value = 0.0004

Conclusion: Since 0.0004 is less than any reasonable α−level we reject $\mathbf{H_0}$. There is sufficient evidence to conclude that the independent variable X_1 is useful for predicting Y.

(b) The 95% confidence interval for β_2 is based on

$$\hat{\beta}_2 \pm t(n - p - 1, \alpha/2)\hat{\sigma}_{\hat{\beta}_2}.$$

Since $t(5, 0.025) = 2.571$, the confidence interval becomes

$$-0.0788 \pm 2.571(0.0371)$$

or (-0.1744,0.0168). Thus, at the 95% confidence level, we can conclude that the change in Y as X_2 increases by 1 unit can be anywhere from a decrease of 0.1744 to an increase of 0.0168.

(c) The 99% confidence interval for $\theta = \beta_1 - \beta_2$ is based on

$$\hat{\theta} \pm t(n - p - 1, \alpha/2)\hat{\sigma}_{\hat{\theta}}.$$

The constants are $c_0 = 0, c_1 = +1$, and $c_2 = -1$ so that

$$\hat{\theta} = \hat{\beta}_1 - \hat{\beta}_2 = 0.8000 - 0.0788 = 0.7212.$$

Then, since $s_{12} = s_{21}$,

$$\hat{\sigma}_{\hat{\theta}}^2 = \hat{\sigma}^2 \left(c_1^2 s_{11} + c_2^2 s_{22} + 2c_1 c_2 s_{12} \right) = 0.0637(0.1465 + 0.0217 + 2(0.0812)) = 0.0210.$$

Thus the confidence interval is

$$0.7212 \pm 4.032\sqrt{0.0210}$$

or (0.1369,1.3055). Thus, at the 95% confidence level, a range of plausible values for $\beta_1 - \beta_2$ is (0.3486,1.0938). Since this interval does not contain zero, there is sufficient evidence to say that the two slopes are different ($p \leq 0.05$).

(d) The estimate of the expected response at $x_o = (1, 5, 3)$ is

$$\hat{\mu}(x_o) = -0.8611 + 0.8000(5) + -0.0788(3) = 2.872.$$

The variance of this estimate is

$$\hat{\sigma}^2(x_o) = \hat{\sigma}^2 \left\{ \frac{1}{n} + (x_{01} - \bar{x}_1)^2 s_{11} + (x_{02} - \bar{x}_2)^2 s_{22} + 2(x_{01} - \bar{x}_1)(x_{02} - \bar{x}_2)s_{12} \right\}$$

$$= 0.0637 \left\{ \frac{1}{8} + (5 - 4.9625)^2 \, 0.1464 + (3 - 5.9875)^2 \, 0.0217 \right.$$

$$\left. -2(5 - 4.9625)(3 - 5.9875)0.0098 \right\}$$

$$= 0.0204$$

Then the 95% lower confidence bound for the expected response at $X_1 = 5$ and $X_2 = 3$ is

$$\hat{\mu}(x_o) - t(n - p - 1, \alpha)\hat{\sigma}^2 \hat{\mu}(x_o)$$

$= 2.8725 - 2.015\sqrt{0.0204} = 2.5847$. Thus, at the 95% confidence level, the expected response at $X_1 = 5$ and $X_2 = 3$ is at least 2.5847.

(e) The 95% confidence interval for an observed response is based on the same point estimate as above. The variance formula is slightly different. As stated in the text it is (with $m = 1$)

$$\hat{\sigma}_p^2(x_o) = \hat{\sigma}^2 \left\{ 1 + \frac{1}{n} + (x_{01} - \bar{x}_1)^2 s_{11} + (x_{02} - \bar{x}_2)^2 s_{22} + 2(x_{01} - \bar{x}_1)(x_{02} - \bar{x}_2)s_{12} \right\}$$

$$= 0.0637 \left\{ 1 + \frac{1}{8} + (5 - 4.9625)^2 0.1464 + (3 - 5.9875)^2 0.0217 \right.$$

$$\left. -2(5 - 4.9625)(3 - 5.9875)0.0098 \right\}$$

$$= 0.0841.$$

Then the 95% prediction interval for an observed response when $X_1 = 5$ and $X_2 = 3$ is $2.8725 \pm 2.571\sqrt{0.0841} = (2.1269, 3.6181)$. Thus, at the 95% confidence level, it is plausible that an observed response would be between 2.1269 and 3.6181 when $X_1 = 5$ and $X_2 = 3$.

(f) The interval in Part Part e in Exercise **11.17** will be wider than the corresponding 95% confidence interval for the expected response at $X_1 = 5$ and $X_2 = 3$. This is because the expected response does not involve the random error component. A particular observed response, however, is equal to the expected response plus random error. The prediction interval takes this into account by including the variability of the random error component in its standard error.

(g) In Model 2, the change in the expected response resulting from a 1 unit increase in X_1 will be the same at all values of X_2 only if $\beta_3 = 0$. Hence, to answer the question we can perform the following test.

H$_0$: $\beta_3 = 0$

H$_1$: $\beta_3 \neq 0$

Test Statistic: $p-$value $= 0.8582$

Conclusion: Since 0.8582 is larg, there is not sufficient evidence to say that the change in the expected response resulting from a 1 unit increase in X_1 is different at all values of X_2. Hence, there is no evidence of an interaction between X_1 and X_2.

11.19. (a) The standard error for $\hat{\beta}_i$ is $\hat{\sigma}_{\hat{\beta}_i} = \sqrt{\hat{\sigma}^2 s_{ii}}$. $SS[E]$ in Example 11.9 was 4.2614. Thus

$\hat{\sigma}^2 = 4.2614/4 = 1.0654$ and the standard errors are

$$\hat{\sigma}_{\hat{\beta}_0} = \sqrt{(1.0654)(0.8179)} = 0.9335$$

$$\hat{\sigma}_{\hat{\beta}_1} = \sqrt{(1.0654)(18.864)} = 4.4829$$

$$\hat{\sigma}_{\hat{\beta}_2} = \sqrt{(1.0654)(17.252)} = 4.2871$$

(b) **H_0: $\beta_1 = 0$**
  H_1: $\beta_1 \neq 0$
  Test Statistic: $t_c = \dfrac{51.2610}{4.4829} = 11.434$
  Rejection Region: $t(n - p - 1, \alpha/2) = t(4, 0.025) = 2.776$
  Reject H_0 if
  $t_c < -2.776$ or $t > 2.776$
  Conclusion: Since 11.434 is greater than 2.776, we reject H_0. There is sufficient evidence at the 95% confidence level to say that $\beta_1 \neq 0$.

H_0: $\beta_2 = 0$
  H_1: $\beta_2 \neq 0$
  Test Statistic: $t_c = \dfrac{-28.5775}{4.2871} = -6.67$
  Rejection Region: $t(n - p - 1, \alpha/2) = t(4, 0.025) = 2.776$
  Reject H_0 if
  $t_c < -2.776$ or $t > 2.776$
  Conclusion: Since -6.67 is less than -2.776, we reject H_0. There is sufficient evidence at the 95% confidence level to say that $\beta_2 \neq 0$.

Since both of these tests rejected H_0, we conclude that the quadratic model in concentration of the enzyme is useful for predicting the adsorbed amount of enzyme.

(c) The 95% confidence bounds for the estimate of $\mu'(c)$ is calculated with matrix formulation as

$$\hat{\theta} = \hat{\mu}'(c) = \hat{\beta}_1 + 2\hat{\beta}_2 c = 51.261 + 2(-28.5774)c$$

The standard error of $\hat{\theta}$ is

$$\sigma_{\hat{\theta}} = \sigma x_0' S x_0.$$

For $c = 0.25$, this becomes

$$\hat{\mu}'(0.25) = 51.261 + 2(-28.5774)(0.25)$$
$$= 36.97230$$

$$\sigma_{\hat{\theta}} = \sqrt{1.06535}\sqrt{\begin{bmatrix} 1 & 0.25 \end{bmatrix} \begin{bmatrix} 18.864 & -17.39 \\ -17.39 & 17.252 \end{bmatrix} \begin{bmatrix} 1 \\ 0.25 \end{bmatrix}}$$

$$= 3.46154$$

The 95% confidence interval for $\mu'(0.25)$ is then

$$36.97230 \pm t(4, 0.025)(3.46154)$$

or $(30.07955, 43.86505)$.

The figure shows the entire 95% confidence bands for $\mu'(c)$.

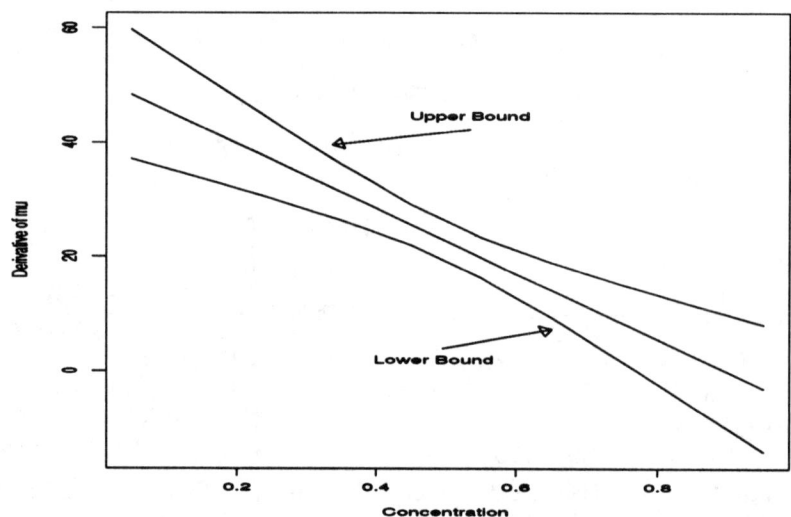

This band estimates the rate at which the expected adsorption changes with increasing enzyme concentration. For example, at the 95% confidence level the rate of change of the expected adsorption could be as much as 51.74 or as little as 33.63 when the increase in enzyme concentration is 0.15. It is clear from the figure that the rate of change of the expected adsorption decreases with increasing concentration.

11.21. With $c = \begin{bmatrix} 0 \\ .6 \\ .2 \end{bmatrix}$ the matrix representation given in the text becomes

$$\hat{\sigma}_{\hat{\theta}}^2 = 0.0930 \begin{bmatrix} 0 & .6 & .2 \end{bmatrix} \begin{bmatrix} 1.4459 & -0.5577 & -4.4753 \\ -0.5577 & 0.5868 & 0.8462 \\ -4.7453 & 0.8463 & 19.1588 \end{bmatrix} \begin{bmatrix} 0 \\ .6 \\ .2 \end{bmatrix}$$

$$= 0.0930 \begin{bmatrix} -1.2837 & 0.5213 & 4.3395 \end{bmatrix} \begin{bmatrix} 0 \\ .6 \\ .2 \end{bmatrix} = 0.0930(1.1807) = 0.1098.$$

Then $\hat{\sigma}_{\hat{\theta}} = \sqrt{0.1098} = 0.3314$ and the lower bound becomes

$$\hat{\theta}_L = \hat{\theta} - t(17, .10)(0.3314) = 1.4348 - (1.333)(0.3314) = 0.9930$$

which is the value obtained in Example 11.15.

11.23. First off, if $\beta_1 - \beta_3 = 0$, then $\beta_1 = \beta_3$. Similarly, if $\beta_2 - \beta_3 = 0$, then $\beta_2 = \beta_3$. Since β_1 and β_2 both equal β_3, it follows that $\beta_1 = \beta_2$. From this (11.21) follows.

Going in the other direction, $\beta_1 = \beta_2 = \beta_3$ clearly implies that $\beta_1 = \beta_3$ which implies that $\beta_1 - \beta_3 = 0$. Also, $\beta_1 = \beta_2 = \beta_3$ clearly implies that $\beta_2 = \beta_3$ which, in turn, implies $\beta_2 - \beta_3 = 0$.

Another set of restrictions would be $\beta_1 - \beta_2 = 0$ and $\beta_2 - \beta_3 = 0$.

11.25. (a) Since the total sum of squares does not depend upon the assumed model, the total sum of squares for the complete quadratic model is the same as the total sum of squares for the two models shown in Exercise **11.12**, i.e. 608995.04. Then
$SS[R] = SS[TOT] - SS[E] = 608995.04 - 339971.7162 = 269023.3238.$

120

(b) The ANOVA table associated with the quadratic model is

Source	df	SS	MS
Model	5	269,023.3238	53,804.6648
Error	19	339,971.7162	17,893.2482
C Total	24	608995.04	

(c) H_0: $\beta_1 = \beta_2 = \beta_3 = \beta_4 = \beta_5 = 0$

H_1: At least one $\beta_i \neq 0$, for $i = 1, 2, 3, 4, 5$

Test Statistic: $F_c = 53804.6648/17893.2482 = 3.0070$

Comparing this with the F(5,19) distribution, we have $0.05 > p-\text{value} > 0.025$.

Conclusion: Whether we reject or fail to reject the null hypothesis depends upon the desired level of significance. At the 5% level of significance, we would reject the null hypothesis, but at the 2.5% level, we do not reject the null hypothesis.

(d) $SS[E]_r$ is the $SS[E]$ from the reduced model, that is Model 1 from Exercise **11.12**. $SS[E]_f$ is the $SS[E]$ from the full model, that is the quadratic model whose ANOVA table is specified above. We then have
$R(\beta_3, \beta_4, \beta_5 | \beta_0, \beta_1, \beta_2) = SS[E]_r - SS[E]_f = 427625.9661 - 339971.7162 = 87654.2499$.
This number is the additional variation in the response that can be explained by adding $x_1 x_2, x_1^2$ and x_2^2 to a model containing x_1 and x_2.

(e) H_0: $\beta_3 = \beta_4 = \beta_5 = 0$

H_1: At least one $\beta_i \neq 0$ for $i = 3, 4, 5$

Test Statistic: $F_c = \dfrac{SS[H_0]/(p-q)}{MS[E]_f} = \dfrac{87654.2499/3}{17893.2482} = 1.6329$

Comparing this with the F(3,19)-distribution, we find $p-\text{value} > 0.10$.

Conclusion: Since the $p-$value is large we do not reject H_0. Thus, there is not sufficient evidence to indicate that the second order terms are useful for predicting the stress of hospitalized patients given that the first order terms are in the model.

(f) $R(\beta_4, \beta_5 | \beta_0, \beta_1, \beta_2, \beta_3) = 347831.10194 - 339971.7162 = 7859.3857$. This number is the additional variation in the response that can be explained by adding x_1^2 and x_2^2 to a model containing $x_1 x_2, x_1$ and x_2.

(g) H_0: $\beta_4 = \beta_5 = 0$

H_1: At least one $\beta_i \neq 0$ for $i = 4, 5$

Test Statistic: $F_c = \dfrac{7859.3857/2}{17893.2482} = 0.2196$

Comparing this with the F(2,19)-distribution, we find $p-\text{value} > 0.10$

Conclusion: Since the $p-$value is large we do not reject H_0. Thus, there is not sufficient evidence to indicate that the quadratic terms are useful for predicting the stress of hospitalized patients given that the first order terms and the interaction term are in the model.

11.27. (a) $s_{obs}^2 = 0.23067$ and $s_{pred}^2 = 0.1898$.

(b) Using the ratio of the variances of the observed and predicted values, the sample coefficient of determination is

$$R_{Y.X_1, X_2, X_3}^2 = \frac{0.1898}{0.23067} = 0.8228.$$

(c) First, note that $s_{xy}^2 = 0.1898$. Then, the simple correlation coefficient as defined in Chapter 10 is

$$r_{XY} = \frac{s_{xy}}{s_x s_y} = \frac{0.1898}{\sqrt{(0.23067)(0.1898)}} = 0.9071.$$

The multiple correlation coefficient is the square root of the coefficient of determination or $R_{Y.X_1, X_2, X_3} = \sqrt{0.8228} = 0.9071$ which is exactly the same as the simple correlation coefficient.

(d) The correlation coefficient for the given prediction equation will be less than the the correlation coefficient calculated in Part Part c in Exercise **11.27**. The correlation coefficient in Part Part c in Exercise **11.27** is based on the least squares equation. By definition this function fits the data best, which means that the predicted values are as close to the observed values as they can be. The predicted values from any other equation, and in particular from the one given, will, in general, be farther from the observed values than the least squares predicted values. The correlation coefficient based on the least squares estimates will be larger than the correlation coefficient based on any other line.

(e) The partial correlation coefficient between log(leafburn time) and percent nitrogen after adjusting for percent chlorine and percent potassium is the positive square root of the appropriate partial coefficient of determination. From the ANOVA table in Example 11.7.4 we can get the necessary Type I and Type III Sum of Squares. Then the partial coefficient of determination is

$$R_{YX_3.X_1X_2}^2 = \frac{R(\beta_1|\beta_0, \beta_2, \beta_3)}{SS[TOT] - R(\beta_2, \beta_3|\beta_0)}$$

But

$$\begin{aligned}
R(\beta_2, \beta_3|\beta_0) &= R(\beta_1, \beta_2, \beta_3|\beta_0) - R(\beta_1|\beta_0, \beta_2, \beta_3) \\
&= R(\beta_3|\beta_0, \beta_1, \beta_2) + R(\beta_1, \beta_2|\beta_0) - R(\beta_1|\beta_0, \beta_2, \beta_3) \\
&= R(\beta_3|\beta_0, \beta_1, \beta_2) + R(\beta_1|\beta_0) + R(\beta_2|\beta_0, \beta_1) - R(\beta_1|\beta_0, \beta_2, \beta_3) \\
&= 1.2049 + 3.4460 + 0.8538 - 2.6587 \\
&= 2.846
\end{aligned}$$

Now we can calculate

$$R_{YX_3.X_1X_2}^2 = \frac{2.6587}{6.68952 - 2.846} = 0.6917.$$

Then $R_{YX_3.X_1X_2} = +\sqrt{0.6917} = 0.8317$. The positive sign is used because the slope is positive and the sign of the two are the same.

(f) There is a fairly strong linear association between the portions of log(leafburn time) and percent nitrogen left unexplained by percent chlorine and percent potassium. It appears that additional useful information about the response can be gained by the incorporation of a linear term in percent nitrogen into the model.

11.29. (a) The figure shows a Normal(q-q) plot and a plot of the residuals versus the predicted values. These two plots can be used to see if the assumption of independent $N(0, \sigma^2)$ errors is reasonable for this data set.

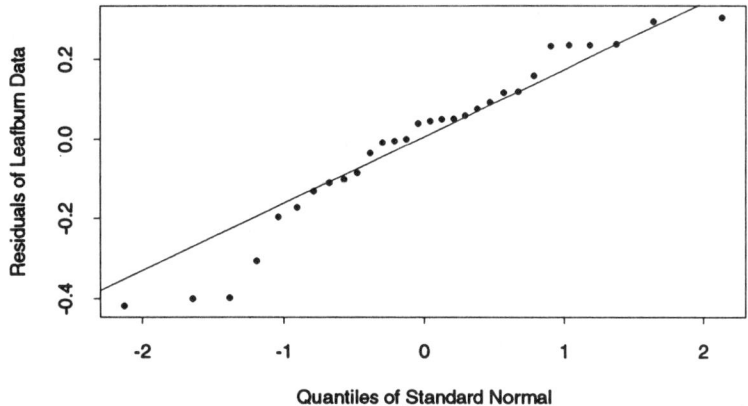

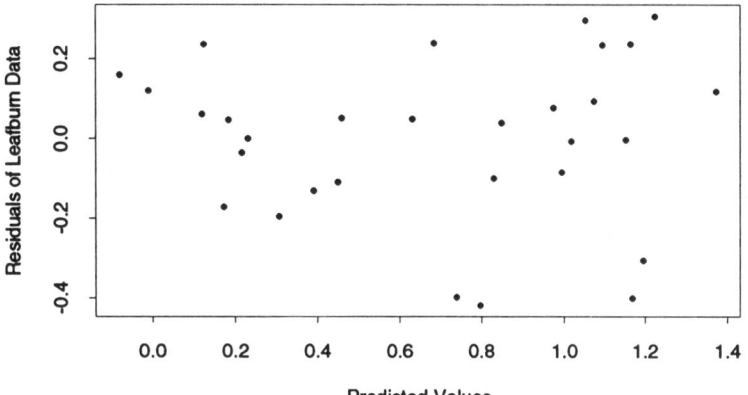

The Normal(q-q) plot shows that there might two places of concern. The left tail of the data appears to be longer than the tail for the Standard Normal distribution. This is evidenced by the data points being much lower than the straight line. There also seems to be a bunching in the right tail, but this evens out as it gets to the end of the data.

The plot of residuals versus the predicted values might be showing a larger variance for higher predicted values. Below a predicted value of 0.7, the range of the residuals is roughly -0.2 to 0.2. Above 0.7, the range increases to roughly -0.4 to 0.3. This might warrant some further investigation.

(b) Given the appropriate sum of squares, R^2 and C_p can be calculated as

$$R^2 = 1 - \frac{SS[E]}{SS[TOT]}$$

and

$$C_p = p + 1 + \frac{(MS[E:p] - MS[E:m])(n-p-1)}{MS[E:m]}.$$

The ANOVA tables for the three subsets of interest are calculated from the cumulative Type I Sum of Squares. R^2 and C_p can be calculated from the ANOVA tables. This information is summarized below.

	Source	df	SS	MS	R^2	C_p
	Model	1	3.446	3.446	0.515	45.1767
$\{X_1\}$	Error	28	3.2435	0.1158		
	C Total	29	6.6895			

	Source	df	SS	MS	R^2	C_p
	Model	2	4.2998	2.1499	0.6428	28.4358
$\{X_1, X_2\}$	Error	27	2.3897	0.0885		
	C Total	29	6.6895			

	Source	df	SS	MS	R^2	C_p
	Model	3	5.5047	1.8349	0.7238	4
$\{X_1, X_2, X_3\}$	Error	26	1.8479	0.0456		
	C Total	29	6.6895			

The subsets $\{X_1\}$ and $\{X_1, X_2\}$ are not adequate for explaining log(leafburn time). They each have low R^2 values (0.515 and 0.6428) and high C_p values (45.1767 and 28.4358). The third subset, $\{X_1, X_2, X_3\}$ has the highest R^2 value (0.7238) and a C_p value which is exactly $p + 1 = 4$. Therefore, $\{X_1, X_2, X_3\}$ is the "best" subset to use as a model for explaining log(leafburn time).

Chapter 12

The General Linear Model

12.1. The first hypothesis is
$$\beta_1 = \beta_2 = \cdots = \beta_{t-1} = 0.$$

Substituting the first equivalence given in Equation (12.8) into this hypothesis, it becomes
$$\tau_1 - \tau_t = \tau_2 - \tau_t = \cdots = \tau_{t-1} - \tau_t = 0.$$

Then, adding τ_t to both sides of each equality gives
$$\tau_1 = \tau_2 = \tau_3 = \cdots = \tau_t.$$

Now, starting with
$$\tau_1 = \tau_2 = \tau_3 = \cdots = \tau_t,$$
subtracting τ_t from both sides of each equality gives
$$\tau_1 - \tau_t = \tau_2 - \tau_t = \cdots = \tau_{t-1} - \tau_t = 0.$$

But, by Equation (12.8) $\beta_i = \tau_i - \tau_t$. Using this relationship the above becomes
$$\beta_1 = \beta_2 = \cdots = \beta_{t-1} = 0.$$

This shows that
$$H_{0\beta} \iff H_{0\tau}.$$

Finally, start with
$$\tau_1 = \tau_2 = \tau_3 = \cdots = \tau_t.$$

If μ, the overall mean, is added to each τ_i and knowing from Equation (12.8) that $\mu_i = \mu + \tau_i$ we have
$$\mu_1 = \mu_2 = \mu_3 \cdots = \mu_t.$$

Working the other direction, we start with
$$\mu_1 = \mu_2 = \mu_3 \cdots = \mu_t$$

and subtract the overall mean from both sides of each equality. This gives
$$\mu_1 - \mu = \mu_2 - \mu = \cdots = \mu_t - \mu.$$

But, from Equation (12.8), $\mu_i = \mu + \tau_i$ or $\tau_i = \mu_i - \mu$. Substituting these into the above equalities gives
$$\tau_1 = \tau_2 = \tau_3 = \cdots = \tau_t.$$

to show the equivalence
$$H_{0\tau} \iff H_{0\mu}.$$

Hence, the three null hypotheses are equivalent to each other.

12.3. (a) The sample statistics for the three treatments are

Trt (i)	1	2	3
$\bar{x}_i$	52.8	35.6	37.8
s_i^2	153.9556	64.0444	173.9556

and the overall mean is 42.0667. Then, the sums of squares are

$$SS[TRT] = \sum \frac{Y_{i+}^{2}}{n_i} - CM$$
$$= 1752.2667$$

and

$$SS[E] = \frac{(n_1 - 1)s_1^2 + (n_2 - 1)s_2^2 + (n_3 - 1)s_3^2}{n_1 + n_2 + n_3 - 3}$$
$$= \frac{(9)(153.9556) + (9)(64.0444) + (9)(173.9556)}{27}$$
$$= 3527.6000$$

Then the Analysis of Variance table is

Source	df	SS	MS	F
Treatments	2	1752.2667	876.1333	6.706
Error	27	3527.6000	130.6519	
Total	29	5279.8667		

(b) The ANOVA table obtained in Part Part **a** in Exercise **12.3** is identical (after allowing for roundoff errors) to the one given in Display 12.1. This shows that the ANOVA model and the regression model are identical for this type of data.

12.5. (a) A general linear model for the plant height data would be

$$Y_i = \beta_0 + \beta_1 x_{i1} + \beta_2 x_{i2} + \beta_3 x_{i3} + \beta_4 x_{i4} + E_i, \qquad i = 1, \ldots, 20,$$

where

Y_i is the measured height for a plant with $X_1 = x_{i1}$, $X_2 = x_{i2}$, $X_3 = x_{i3}$, $X_4 = x_{i4}$,

$X_1 = \begin{cases} 1 & \text{if the response is for treatment D} \\ 0 & \text{otherwise} \end{cases}$

$X_2 = \begin{cases} 1 & \text{if the response is for treatment AL} \\ 0 & \text{otherwise} \end{cases}$

$X_3 = \begin{cases} 1 & \text{if the response is for treatment AH} \\ 0 & \text{otherwise} \end{cases}$

$X_4 = \begin{cases} 1 & \text{if the response is for treatment BL} \\ 0 & \text{otherwise} \end{cases}$

E_i is the random error associated with Y_i,

(b) The parameters are interpreted as

$\beta_0 = \mu_5 =$ the expected response of treatment BH,

$\beta_1 = \mu_1 - \mu_5 =$ the difference between the expected responses of treatment D and treatment BH,

$\beta_2 = \mu_2 - \mu_5 =$ the difference between the expected responses of treatment AL and treatment BH,

$\beta_3 = \mu_3 - \mu_5 =$ the difference between the expected responses of treatment AH and treatment BH,

$\beta_4 = \mu_4 - \mu_5 =$ the difference between the expected responses of treatment BL and treatment BH.

(c) The estimates for the expected responses of each treatment are the sample means of each treatment given in Table 8.2. The population regression parameters are functions of the population treatment means as shown in Part b in Exercise **12.5**. The estimates of the regression parameters are given by taking these same functions of the sample means. For example, β_1 equals $\mu_1 - \mu_5$. Thus, the estimate of β_1 is $\bar{x}_1 - \bar{x}_5 = 34.02 - 34.29 = -0.27$.

The estimated standard errors are given in Box 4.4C. These are $\sqrt{1.09\left(\frac{1}{4}\right)} = 0.522$ for β_0 and $\sqrt{1.09\left(\frac{1}{4} + \frac{1}{4}\right)} = 0.7382$ for all of the other parameters. The test statistic is simply the ratio of the parameter estimate to the standard error. This information gives the following table

	Function of Treatment Means	Parameter Estimates	Estimated Std Errors	Test Statistic
β_0	μ_5	$34.29 = 34.29$	0.5220	65.6897
β_1	$\mu_1 - \mu_5$	$34.02 - 34.29 = -0.27$	0.7382	-0.3657
β_2	$\mu_2 - \mu_5$	$31.90 - 34.29 = -2.39$	0.7382	-3.2376
β_3	$\mu_3 - \mu_5$	$30.84 - 34.29 = -3.45$	0.7382	-4.6735
β_4	$\mu_4 - \mu_5$	$34.31 - 34.29 = 0.02$	0.7382	0.0271

(d) At the $\alpha = 0.05$ significance level, the critical value for these tests is $t(15, 0.025) = 2.131$. Using this critical value, the parameters β_0, β_2 and β_3 are statistically significantly different from zero. For each of the other parameters, at the .05 level, we cannot say that the parameter is significantly different from zero.

Interpreting these results in terms of the expected responses for the four safe-lights, we can make the following statements. At the .05 level there is sufficient evidence to say that the mean response for treatment BH is different from zero. At the same level, there is not sufficient evidence to say that the mean response for treatment D is different from the mean response for treatment BH. At the .05 level, there is sufficient evidence to say that the mean response for treatment AL is different from the mean response for treatment BH. At the .05 level, there is sufficient evidence to say that the mean response for treatment AH is different from the mean response for treatment BH. Finally, at the .05 level, there is not sufficient evidence to say that the mean response for treatment BL is different from the mean response for treatment BH.

12.7. (a) Since there are two treatments, high and normal phosphate, there will be one dummy independent variable. The model is

$$Y_i = \beta_0 + \beta_1 x_i + E_i, \qquad i = 1, \dots 24$$

where

Y_i is the frequency of painful crises for the ith patient,

$X_i = \begin{cases} 0 & \text{if phosphate level is high} \\ 1 & \text{if phosphate level is normal} \end{cases}$

(b) The "slope parameter", β_1, measures the difference in the expected frequencies for patients with normal phosphate and patients with high phosphate. The estimate of this is the difference in sample treatment means. The intercept is the expected response for patients with high phosphate. For the model above, the estimate of the intercept is the sample mean for the patients with high phosphate. This is 0.218. Then the estimate of the "slope parameter" is $0.0607 - 0.2180 = -0.1573$.

The 90% lower confidence bound for β_1 is $\beta_1 - t(22, 0.10)\hat{\sigma}_{\hat{\beta}_1}$. The S matrix can be calculated as

$$S = \begin{bmatrix} \dfrac{1}{n} + \dfrac{\bar{x}^2}{S_{xx}} & -\dfrac{\bar{x}}{S_{xx}} \\ -\dfrac{\bar{x}}{S_{xx}} & \dfrac{1}{S_{xx}} \end{bmatrix}$$

With these calculations the S matrix is

	INTERCEPT	X
INTERCEPT	0.1	−0.1
X	−0.1	0.1714

Further, $\hat{\sigma}^2 = 0.0049$, $c_0 = 0$, and $c_1 = 1$. This information can be used to calculate the standard error of the slope,

$$\hat{\sigma}_{\hat{\beta}_1} = \sqrt{0.0049(0.1714)} = 0.0290.$$

Now, $t(22, 0.10) = 1.321$, so that the lower bound becomes

$$-0.1573 - 1.321(0.0290) = -0.1956.$$

(c) The lower bound calculated above is -0.1956. Thus, with 90% confidence, we can conclude that the expected frequency of painful crises for patients with normal phosphate is at most 0.1956 crises per month lower than for patients with higher phosphate.

12.9. (a) The general linear model would be

$$Y_i = \beta_0^* + \beta_1^* x_{i1} + \beta_2^* x_{i2} + E_i, \qquad i = 1, \ldots, 30$$

where

Y_i is the post-test score for the ith subject observed at the value $X_1 = x_{i1}$ and $X_2 = x_{i2}$,

$$X_1 = \begin{cases} 0 & \text{if Treatment 1} \\ 1 & \text{if Treatment 2} \\ -1 & \text{if Treatment 3} \end{cases},$$

$$X_2 = \begin{cases} 1 & \text{if Treatment 1} \\ 0 & \text{if Treatment 2} \\ -1 & \text{if Treatment 3} \end{cases},$$

E_i the random error for the ith subject.

(b) The combinations of parameters from the above model that equal the expected response for each treatment are

$$\begin{aligned} \mu_1 &= \beta_0^* + \beta_2^* \\ \mu_2 &= \beta_0^* + \beta_1^* \\ \mu_3 &= \beta_0^* - (\beta_1^* + \beta_2^*) \end{aligned}$$

The parameters from the models at Equations 12.1 and 12.2 as related to the expected response for each treatment are

	ANOVA	Regression
μ_1	$\mu + \tau_1$	$\beta_0 + \beta_1$
μ_2	$\mu + \tau_2$	$\beta_0 + \beta_2$
μ_3	$\mu + \tau_3$	β_0

128

Thus, the "old" parameters are interpreted as

μ is the overall mean value of the expected response,

τ_1 is the effect on the expected response due to treatment 1,

τ_2 is the effect on the expected response due to treatment 2,

τ_3 is the effect on the expected response due to treatment 3,

β_0 is the expected response of treatment 3,

β_1 is the difference in expected response between treatment 1 and treatment 3, and

β_2 is the difference in expected response between treatment 2 and treatment 3.

Now, the interpretations of the parameters in the new formulation of the model are

$\beta_0^* = \frac{\mu_1+\mu_2+\mu_3}{3} = \bar{\mu}$, say,

$\beta_1^* = \mu_2 - \bar{\mu}$, and

$\beta_2^* = \mu_1 - \bar{\mu}$.

So that,

β_0^* is the overall mean value of the expected response,

β_1^* is the deviation of the expected response due to treatment 2 from the overall mean, and

β_2^* is the deviation of the expected response due to treatment 1 from the overall mean

(c) The estimates of the new parameters can be obtained by linear combinations of the estimates of the old parameters, i.e. those given in the text. These linear combinations are

$$\hat{\beta}_0^* = \frac{(\hat{\beta}_0 + \hat{\beta}_1) + (\hat{\beta}_0 + \hat{\beta}_2) + \hat{\beta}_0}{3}$$

$$= \hat{\beta}_0 + \frac{1}{3}\left(\hat{\beta}_1 + \hat{\beta}_2\right),$$

$$\hat{\beta}_1^* = \left(\hat{\beta}_0 + \hat{\beta}_2\right) - \left\{\hat{\beta}_0 + \frac{1}{3}\left(\hat{\beta}_1 + \hat{\beta}_2\right)\right\}$$

$$= \frac{2}{3}\hat{\beta}_2 - \frac{1}{3}\hat{\beta}_1$$

$$\hat{\beta}_2^* = \left(\hat{\beta}_0 + \hat{\beta}_1\right) - \left\{\hat{\beta}_0 + \frac{1}{3}\left(\hat{\beta}_1 + \hat{\beta}_2\right)\right\}$$

$$= \frac{2}{3}\hat{\beta}_1 - \frac{1}{3}\hat{\beta}_2$$

Using the information in Part 2 of the SAS output we can calculate how the entries in Part 2 wwould change when the GLM procedure of SAS is used to fit the model in Part Part a in Exercise **12.9** to the mental capacity data. First, the estimates and their

129

standard errors are

$$\hat{\beta}_0^* = 37.8 + \frac{1}{3}(15.0 + -2.2) = 42.0667,$$

$$\hat{\sigma}_{\hat{\beta}_0^*} = \sqrt{\hat{\sigma}_{\hat{\beta}_0}^2 + \frac{1}{9}\left(\hat{\sigma}_{\hat{\beta}_1}^2 + \hat{\sigma}_{\hat{\beta}_2}^2\right)} = \sqrt{3.6146^2 + \frac{1}{9}(5.1118^2 + 5.1118^2)} = 4.3442;$$

$$\hat{\beta}_1^* = \left(\frac{2}{3}\right)(-2.2) - \left(\frac{1}{3}\right)(15.0) = -6.4667,$$

$$\hat{\sigma}_{\hat{\beta}_1^*} = \sqrt{\frac{4}{9}\hat{\sigma}_{\hat{\beta}_2}^2 + \frac{1}{9}\hat{\sigma}_{\hat{\beta}_1}^2} = \sqrt{\left(\frac{4}{9}\right)5.1118^2 + \left(\frac{1}{9}\right)5.1118^2} = 3.8101;$$

$$\hat{\beta}_2^* = \left(\frac{2}{3}\right)(15.0) - \left(\frac{1}{3}\right)(-2.2) = 9.2667,$$

$$\hat{\sigma}_{\hat{\beta}_2^*}^2 = \sqrt{\frac{4}{9}\hat{\sigma}_{\hat{\beta}_1}^2 + \frac{1}{9}\hat{\sigma}_{\hat{\beta}_2}^2} = \sqrt{\left(\frac{4}{9}\right)5.1118^2 + \left(\frac{1}{9}\right)5.1118^2} = 3.8101.$$

With this information we calculate the test statistics and corresponding p-values (calculated in S-plus)

$$t_{\hat{\beta}_0^*} = \frac{42.0667}{4.3442} = 9.6834,$$

$$p\text{-value} = 1.4 \times 10^{-10};$$

$$t_{\hat{\beta}_0^*} = \frac{-6.4667}{3.8101} = 1.6972$$

$$p\text{-value} = 0.0506;$$

$$t_{\hat{\beta}_0^*} = \frac{9.2667}{3.8101} = 2.4321$$

$$p\text{-value} = 0.0110.$$

12.11. (a) Both the F value in the PRESCR row of Type I sum of squares in Part 2 and the t-value in the PRESCR row in Part 3 of Display 12.2 are testing the null hypothesis H_0: $\beta = 0$. It was shown in Chapter 10 that the test statistic for the F-test is the square of the test statistic for the t-test.

(b) If we understand that the data are being fit to a different model in this display than in Exercise **12.4**, then it will not be a surprise if different answers are obtained. The model affects the results in, at least, three interrelated ways: the point estimate, the standard error, and the information required for the point estimate.

The new model requires knowledge of the pre-score as well as the treatment. This means that there is not just one estimate for each treatment, but rather there is one estimate for each pre-score for each treatment. The point estimate is no longer a single value, but now it is an entire line. The extra, stronger requirement gains stronger results. The model fits the data better as evidenced by the MSE which is an order of magnitude smaller than with the previous model.

(c) We will construct a set of 95% simultaneous confidence intervals for the expected post-test scores for each of the three treatments. From the previous part, it is known that these will actually be confidence bands, since the expected post-test score for each treatment is described by a simple linear regression function. Let the vectors of coefficients for the three treatments be denoted as $c_1'(z) = (1, 1, 0, z)$, $c_2'(z) = (1, 0, 1, z)$, and $c_3'(z) = (1, 0, 0, z)$.

The point estimate for the expected post-test score at a given pretest score is

$$\hat{\theta}_i(z) = c_i'(z)\hat{\beta}$$

130

and the standard error for each estimate is

$$\hat{\sigma}_{\hat{\theta}_i(z)} = \sqrt{c_i'(z)Sc_i(z)\hat{\sigma}^2}$$

From the SAS output, $\hat{\sigma}^2 = 14.6376$ and the least squares estimates for the parameters are $\hat{\beta}' = (1.8008, 17.1442, -0.8458, 1.1285)$. The S matrix is given in Part 4 of the SAS output. The $t-$value is $t(27, 0.025/3) = 2.5524$.

Using this information, the 95% simultaneous confidence intervals for the expected post-test scores for each of the three treatments can be given by the following formula

$$c_i'(z)\hat{\beta} \pm 2.5524\sqrt{c_i'(z)Sc_i(z)\hat{\sigma}^2}.$$

The actual interval depends on the treatment and the pretest score. For example, evaluating this for Treatment 1 and pretest score of 42 (which is in the range of the data for Treatment 1) the interval is

$$66.3423 \pm 2.5524(1.5221)$$

which gives $(62.4573, 70.2273)$. The researcher can be 95% confident that the expected post-test score for a subject who has received Treatment 1 and had a pretest score of 42 is between 62.4573 and 70.2273. The entire set of confidence intervals is plotted in the figure below. Each band was calculated over a common range of the covariates. This requires extrapolation for treatment 1 and treatment 2. If it is believed that the model outside of the range is consistent with the model inside the range, then this should not be a problem.

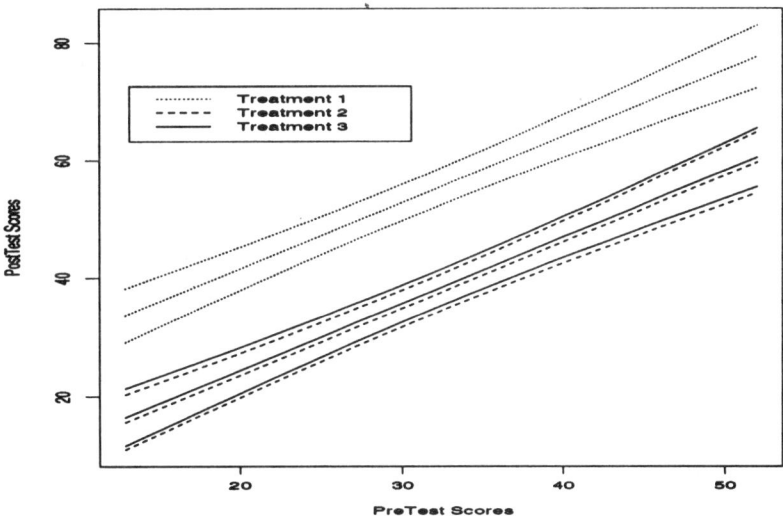

These bands show that there is no evidence of a difference in the expected responses for treatment 2 and treatment 3. However, there is a difference between treatment 1 and the other two treatments. Treatment 1 consistently provides the greater post-test scores for all values of the pretest score.

(d) For treatment 1 and pretest score of 30, the expected post-test score is

$$E(Y) = \beta_0 + \beta_1(1) + \beta_2(0) + \beta_3(30).$$

The estimate for this is

$$\hat{y} = \hat{\beta}_0 + \hat{\beta}_1 + \hat{\beta}_3(30) = 1.7999 + 17.1442 + 1.1285(30) = 52.7991.$$

131

The standard error of the estimate is, using matrix notation,

$$\hat{\sigma}_{\hat{y}} = \sqrt{c'Sc\hat{\sigma}^2} = 1.2098$$

where $c' = (1, 1, 0, 30)$ and S is as in the text.

With this information, the 95% confidence interval is

$$\begin{array}{ccc} \hat{y} & \pm & t(27, 0.025)\hat{\sigma}_y \\ 52.7991 & \pm & 2.5524(1.2098) \end{array}$$

which becomes $(49.7112, 55.8870)$. We can be 95% confident that the expected post-test score of a child whose pretest score is 30 and who received treatment 1 is between 49.7112 and 55.8870.

(e) Using the same matrix notation as above, the formula for the lower bound is

$$c_i'(z)\hat{\beta} - t(27, 0.10/3)\sqrt{1 + c_i'(z)Sc_i(z)\hat{\sigma}^2}.$$

where $t(27, 0.10/3) = 1.9111$. (These formulae can be found in Section 11.5.)
Using this and the vectors $c_1'(30) = (1, 1, 0, 30)$, $c_2'(30) = (1, 0, 1, 30)$, and $c_3'(30) = (1, 0, 0, 30)$, the lower bounds are

$$\begin{array}{rcl} 52.8000 - 1.911(4.0126) & = & 45.1319 \\ 34.8100 - 1.911(4.0130) & = & 27.1412 \text{ and} \\ 35.6558 - 1.911(4.0153) & = & 27.9826. \end{array}$$

With 90% confidence we can expect that the post-test score of a child with pretest score 30 is no lower than 45.1319, if the child received Treatment 1; 27.1412, if the child received Treatment 2; and 27.9826, if the child received Treatment 3.

12.13. (a) The plot of the prediction lines for Diet 1 and Diet 2 as functions of the initial weight is given below

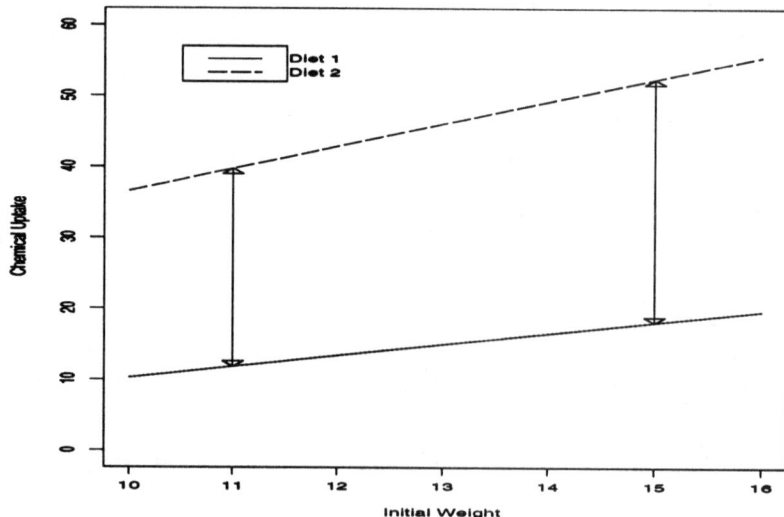

Interaction is when the difference in the expected response between Diet 1 and Diet 2 changes as the initial weight changes. Consider the distance between the two lines

shown in the plot. As the initial weight increases this distance increases, e.g. the distance at initial weight 11 is 27.9456 and at 15 the distance is 34.3376.

The interaction between weight and diet is seen by the change in the distance which, in turn, is caused by the lines not being parallel. If the lines were parallel, then the distance would be constant and there would be no evidence of interaction between weight and diet.

(b) The model used to create the SAS output is

$$Y_j = \beta_0 + \beta_1 x_{1j} + \beta_2 z_j + \beta_3 x_{1j} z_j + E_j$$

where

Y_j denotes the chemical uptake

x_{1j} denotes the type of the diet, and

z_j denotes the initial weight for the jth animal.

The Type I Sum of Squares for diet is denoted as $R(\beta_1|\beta_0)$. The full and reduced model for this sum of squares are

full: $Y_j = \beta_0 + \beta_1 x_{1j} + E_j$

reduced: $Y_j = \beta_0 + E_j$.

The Type III Sum of Squares for diet is denoted by $R(\beta_1|\beta_0, \beta_2, \beta_3)$. The full and reduced model are

full: $Y_j = \beta_0 + \beta_1 x_{1j} + \beta_2 z_j + \beta_3 x_{1j} z_j + E_j$

reduced: $Y_j = \beta_0 + \beta_2 z_j + \beta_3 x_{1j} z_j + E_j$.

The p−value for Type I SS refers to the research hypothesis that diet by itself is a useful predictor of chemical uptake. The Type III SS p−value refers to the research hypothesis that the addition of x_1 to a model containing x_1 and $x_1 z$ improves the model's predictive power.

(c) For an animal with initial weight $Z = z$ and on Diet 1 the expected chemical uptake using the model at (12.6.1) is

$$\mu(1, z) = \beta_0 + \beta_1 + \beta_2 z + \beta_3 z.$$

Similarly, the expected chemical uptake for an animal with initial weight $Z = z$ and on Diet 2 is

$$\mu(0, z) = \beta_0 + \beta_2 z.$$

Thus, the expected difference between the uptake for the two diets (Diet 1 - Diet 2) is

$$\theta(z) = \mu(1, z) - \mu(0, z) = \beta_1 + \beta_3 z.$$

(d) A plot of $\theta(z)$ as a function of z with $\beta_1 = -10$ and $\beta_3 = -2$ is shown in the figure below.

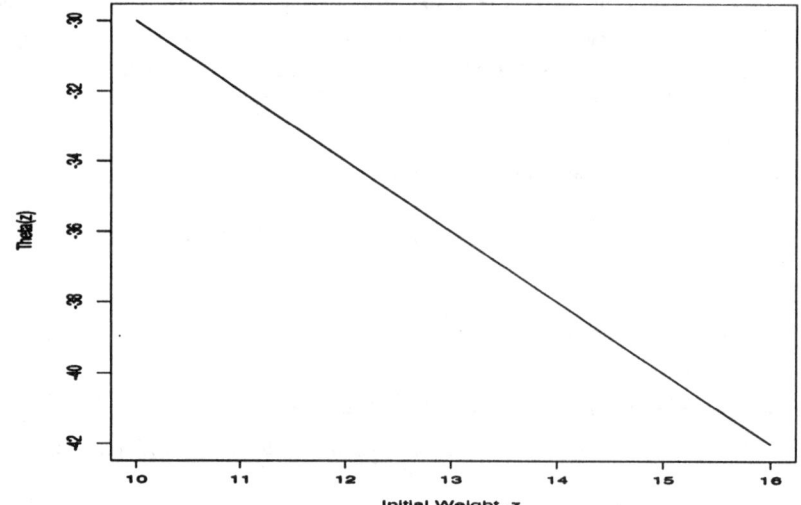

With these values for β_1 and β_3 the expected differential uptake is negative for positive values of Z, indicating that the expected chemical uptake for Diet 2 is greater than for Diet 1. The expected differential uptake increases in magnitude (decreases in value) as the initial weight increases.

(e) $\theta(z) = \beta_1 + \beta_3 z$ can be written as

$$\theta(z) = c_0\beta_0 + c_1\beta_1 + c_2\beta_2 + c_3\beta_3$$

where $c_0 = 0, c_1 = 1, c_2 = 0, c_3 = z$. From Chapter 11

$$\hat{\theta}(z) = c_0\hat{\beta}_0 + c_1\hat{\beta}_1 + c_2\hat{\beta}_2 + c_3\hat{\beta}_3$$

is the unbiased least squares estimate for $\theta(z)$. From the text this is
$\hat{\theta}(z) = -10.3676 - 1.5980z$.

(f) Using the expression in the previous part, we have

$$\begin{aligned}
\hat{\theta}(10) &= -10.3676 - 1.5980(10) = -26.3476 \\
\hat{\theta}(13) &= -10.3676 - 1.5980(13) = -31.1416 \\
\hat{\theta}(16) &= -10.3676 - 1.5980(16) = -35.9356
\end{aligned}$$

(g) The lower bounds for $\theta(z)$ when $Z = 10, 13$, and 16 have point estimates as given in Part f in Exercise **12.13**. The standard errors are

$$\hat{\sigma}_{\hat{\theta}(z)} = \sqrt{c'Sc\hat{\sigma}^2}$$

where $c' = (0, 1, 0, z), \hat{\sigma}^2 = 3.1094$ and S is given in the text. Then the actual standard errors are

$$\begin{aligned}
\hat{\sigma}_{\hat{\theta}(10)} &= 2.0444 \\
\hat{\sigma}_{\hat{\theta}(13)} &= 0.9471 \\
\hat{\sigma}_{\hat{\theta}(16)} &= 1.8215
\end{aligned}$$

The t-value for three 90% simultaneous lower bounds is $t(10, .10/3) = t(10, 0.033) = 2.0576$. Then the three lower bounds are

134

$$z = 10: \qquad -26.3476 - (2.0576)(2.0444) = -30.5533,$$
$$z = 13: \qquad -31.1416 - (2.0576)(0.9471) = -33.0904, \text{ and}$$
$$z = 16: \qquad -35.9356 - (2.0576)(1.8215) = -39.6835.$$

These bounds indicate that we can be 90% confident that the expected difference between uptake for animals on Diet 1 versus animals on Diet 2 is at least -30.5533 when the initial weight is 10, -33.0904 when the the initial weight is 13, and -39.6835 when the initial weight is 16.

In more direct terms, we can be 90% confident that the chemical uptake for animals on Diet 2 is at most 30.5533 more than for animals on Diet 1 when the initial weight is 10, at most 33.0904 when the initial weight is 13, and at most 39.6835 when the initial weight is 16.

12.15. The parameters β_{13} and β_{23} in the model at (12.6.8) measure the effects of the interaction between diet and diet Se concentration and sex and diet Se concentration, respectively. To see this, consider the difference between the models evaluated at diet A and diet B, that is

$$\mu(1, x_2, x_3, x_4) - \mu(0, x_2, x_3, x_4)$$
$$= (\beta_0 + \beta_1 + \beta_2 x_2 + \beta_{12} x_2 + \beta_3 x_3 + \beta_{13} x_3 + \beta_{23} x_2 x_3 + \beta_4 x_4)$$
$$- (\beta_0 + \beta_2 x_2 + \beta_3 x_3 + \beta_{23} x_2 x_3 + \beta_4 x_4)$$
$$= \beta_1 + \beta_{12} x_2 + \beta_{13} x_3.$$

Thus β_{13} measures the change in the expected difference between Diet A and Diet B as x_3, Se concentration, increases by 1 unit. In other words, β_{13} measures the interaction between diet and Se concentration.

Through a similar difference we have

$$\mu(x_1, 1, x_3, x_4) - \mu(x_1, 0, x_3, x_4) = \beta_2 + \beta_{12} x_1 + \beta_{23} x_3$$

and we see that β_{23} measures the expected change in the chemical uptake difference between Males and Females as the diet Se concentration increases, i.e. β_{23} measures the interaction between sex and diet Se concentration.

12.17. (a) Interaction exists between two factors when the differences between the expected responses at any two given levels of one factor will depend upon the value of the other factor. One way, then, to show that the model in the text allows for interaction between age and pre-HbAlc would be to show that the model contains a term that allows for this dependence.

First, let
$$\mu(x_1, x_2, x_3, x_4) = \beta_0 + \beta_1 x_1 + \beta_2 x_2 + \beta_3 x_3 + \beta_4 x_4.$$

Then,
$$\mu(x_1, x_2, 10, x_4) = \beta_0 + \beta_1 x_1 + \beta_2 x_2 + 10\beta_3 + \beta_4 x_4 + 10\beta_{34} x_4$$

is the expected value of Post-HbAlc when $X_1 = x_1, X_2 = x_2, X_3 = 10$, and $X_4 = x_4$. Also,
$$\mu(x_1, x_2, 11, x_4) = \beta_0 + \beta_1 x_1 + \beta_2 x_2 + 11\beta_3 + \beta_4 x_4 + 11\beta_{34} x_4$$

is a similar expected value when $X_3 = 11$. The difference between these two is the difference between the expected responses at two levels of age. This difference is calculated as

$$\mu(x_1, x_2, 10, x_4) - \mu(x_1, x_2, 11, x_4)$$
$$= (\beta_0 + \beta_1 x_1 + \beta_2 x_2 + \beta_3(10) + \beta_4 x_4 + \beta_{34}(10)x_4)$$
$$- (\beta_0 + \beta_1 x_1 + \beta_2 x_2 + \beta_3(11) + \beta_4 x_4 + \beta_{34}(11)x_4)$$
$$= -\beta_3 - \beta_{34} x_4.$$

Since this difference is a function of x_4, pre-HbAlc, the model does allow for interaction between age and pre-HbAlc.

On the other hand, consider the difference between the expected responses at the two levels of sex. First, for females

$$\mu(1, x_2, x_3, x_4) = \beta_0 + \beta_1 + \beta_2 x_2 + \beta_3 x_3 + \beta_4 x_4 + \beta_{34} x_3 x_4$$

and for males

$$\mu(0, x_2, x_3, x_4) = \beta_0 + \beta_2 x_2 + \beta_3 x_3 + \beta_4 x_4 + \beta_{34} x_3 x_4.$$

Then, the difference is

$$
\begin{aligned}
\mu(1, x_2, x_3, x_4) &- \mu(0, x_2, x_3, x_4) \\
&= (\beta_0 + \beta_1 + \beta_2 x_2 + \beta_3 x_3 + \beta_4 x_4 + \beta_{34} x_3 x_4) \\
&- (\beta_0 + \beta_2 x_2 + \beta_3 x_3 + \beta_4 x_4 + \beta_{34} x_3 x_4) \\
&= \beta_1
\end{aligned}
$$

which is a constant and does not depend on the treatment. Therefore, the model does not allow for interaction between sex and treatment.

(b) β_0 is the intercept. It is the expected value of post-HbAlc for males given the standard treatment whose pre-HbAlc is zero.

     β_1 is the partial slope of post-HbAlc on sex. It is the expected difference between the post-HbAlc of females and males.

     β_2 is the partial slope of post-HbAlc on treatment. It is the expected difference between the post-HbAlc of patients receiving the new treatment and those receiving the standard treatment.

     β_3 is the partial slope of post-HbAlc on age. It is the expected change in post-HbAlc as a patient's age increases by one year.

     β_4 is the partial slope of post-HbAlc on pre-HbAlc. It is the expected change in post-HbAlc as the patient's pre-HbAlc increases by one unit.

     β_{34} is the measure of the interaction between age and pre-HbAlc.

(c) First, consider the interaction term, $X_3 * X_4$. Because its p-value is large, 0.4311, we can not say that it is useful for predicting Y. Now we can look at the main effects. X_2 and X_4 have small p-values 0.0017 and 0.0139, respectively, so they are useful for predicting Y. X_1 has an associated p-value of 0.0958 and might be useful for predicting Y. Finally, X_3 has a large associated p-value of 0.3604 and would not be considered significant. However, because X_3 is in the interaction also, one might wonder if there exists collinearity between them. Taking out the interaction effect and re-running the model would be a good next step.

(d) Assuming the data is read in the format given in the text, a SAS program that would produce the information in Display 12.5 could be like the following.

```
DATA BLOOD;
INPUT SUBJECT X3 SEX $ TRT $ X4 Y;
IF SEX = "F" THEN X1 = 1;
ELSE X1 = 0;
IF TRT = "N" THEN X2 = 1;
ELSE X2 = 0;
CARDS;
data inserted here
;
```

136

```
PROC GLM;
MODEL Y = X1 X2 X3 X4 X3*X4;
RUN;
```

12.19. (a) β_0 is the effect of Location 4 on the final weight.

       β_1 is the difference in effects between Location 3 and Location 4.

       β_2 is the difference in effects between Location 2 and Location 4.

       β_{12} is the difference between the sum of effects of Locations 1 and 4 and the sum of the effects of Locations 2 and 3.

       β_3 is the difference in effects between Location 5 and Location 4.

       β is the linear effect due to initial bag weight.

(b) Denoting the parameters defined above as β^n and the parameters in the previous Exercise as β^o, the following are the new parameter estimates.

$$\widehat{\beta}_0^n = \widehat{\beta}_0^o + \widehat{\beta}_4^o = 2.4949 + 1.1071 = 3.602$$

$$\widehat{\beta}_1^n = \widehat{\beta}_0^o + \widehat{\beta}_3^o - (\widehat{\beta}_0^o + \widehat{\beta}_4^o) = \widehat{\beta}_3^o - \widehat{\beta}_4^o = 1.6548 - 1.1071 = .5477$$

$$\widehat{\beta}_2^n = \widehat{\beta}_0^o + \widehat{\beta}_2^o - (\widehat{\beta}_0^o + \widehat{\beta}_4^o) = \widehat{\beta}_2^o - \widehat{\beta}_4^o = -.2803 - 1.1071 = -1.3874$$

$$\widehat{\beta}_{12}^n = \widehat{\beta}_0^o + \widehat{\beta}_1^o - (\widehat{\beta}_0^o + \widehat{\beta}_4^o + (\widehat{\beta}_3^o - \widehat{\beta}_4^o) + (\widehat{\beta}_2^o - \widehat{\beta}_4^o))$$

$$= \widehat{\beta}_1^o - \widehat{\beta}_3^o - \widehat{\beta}_2^o + \widehat{\beta}_4^o = -.2445 - 1.6548 - (-.2803) + 1.1071 = -.5119$$

$$\widehat{\beta}_3^n = \widehat{\beta}_0^o - (\widehat{\beta}_0^o + \widehat{\beta}_4^o) = -\widehat{\beta}_4^o = -1.1071$$

$$\widehat{\beta}^n = \widehat{\beta}_5^o = 1.0832$$

Recalling that $s_{ij} = s_{ji}$ for all i and j, the appropriate standard errors can then be found using the original $\boldsymbol{X'X}^-$ matrix in the usual manner.

$$\hat{\sigma}_{\widehat{\beta}_0^n} = \sqrt{\hat{\sigma}^2(1^2 s_{00} + 1^2 s_{44} + 2s_{04})} = \sqrt{.3016(3.5030 + .7379 + 2(.6297))} = 1.2880$$

$$\hat{\sigma}_{\widehat{\beta}_1^n} = \sqrt{\hat{\sigma}^2(1^2 s_{33} + (-1)^2 s_{44} - 2s_{34})} = \sqrt{.3016(.6115 + .7379 - 2(.4128))} = .3975$$

$$\hat{\sigma}_{\widehat{\beta}_2^n} = \sqrt{\hat{\sigma}^2(1^2 s_{22} + (-1)^2 s_{44} - 2s_{24})} = \sqrt{.3016(.8056 + .7379 - 2(.5196))} = .3900$$

$$\hat{\sigma}_{\widehat{\beta}_{12}^n} = \sqrt{\hat{\sigma}^2(1^2 s_{11} + (-1)^2 s_{33} + (-1)^2 s_{22} + 1^2 s_{44} - 2s_{13} - 2s_{12} + }$$
$$\overline{2s_{14} + 2s_{23} - 2s_{34} - 2s_{24})}$$

$$= \sqrt{.3016(1.1023 + .6115 + .8056 + .7379 - 2(.5091) - 2(.6790) + }$$
$$\overline{2(.6285) + 2(.4345) - 2(.4128) - 2(.5196))} = .5869$$

$$\hat{\sigma}_{\widehat{\beta}_3^n} = \sqrt{\hat{\sigma}^2((-1)^2 s_{44})} = \sqrt{.3016(.7379)} = .4718$$

$$\hat{\sigma}_{\widehat{\beta}} = \sqrt{\hat{\sigma}^2(1^2 s_{55})} = \sqrt{.3016(.0075)} = .0476.$$

(c) The confidence intervals have the general form

$$\hat{\beta}_j \pm t(14, .025)\hat{\sigma}_{\hat{\beta}_j}.$$

For β_0 this becomes

$$3.602 \pm 2.1448(1.2880) \quad \text{or} \quad (.8395, 6.3645).$$

With 95% confidence we can conclude that Location 4 increases the final weight of oysters by between .8395 and 6.3645.

For β_1 this becomes

$$.5477 \pm 2.1448(.3975) \quad \text{or} \quad (-.3049, 1.4003).$$

Since the interval for β_1 contains zero, at the $\alpha = .05$ level, there is insufficient evidence to conclude that Location 3 and 4 have different effects on final weight of oysters.

For β_2 this becomes

$$-1.3874 \pm 2.1448(.3900) \quad \text{or} \quad (-2.2239, -.5509).$$

With 95% confidence we can conclude that the final weight of oysters grown at Location 2 is between 2.2239 and .5509 below the final weight of oysters grown at Location 4.

For β_{12} this becomes

$$-.5119 \pm 2.1448(.5869) \quad \text{or} \quad (-1.7707, .7469).$$

Since the interval for β_{12} contains zero, at the $\alpha = .05$ level, there is insufficient evidence to believe that the sum effect of Locations 1 and 4 on final oyster weight is different than the sum effect of Locations 2 and 3.

For β_3 this becomes

$$-1.1071 \pm 2.1448(.4718) \quad \text{or} \quad (-2.1190, -.0952).$$

With 95% confidence we can conclude that the final weight of oysters grown at Location 5 is between 2.1190 and .0952 below the final weight of oyster grown at Location 4.

For β this becomes

$$1.0832 \pm 2.1448(.0476) \quad \text{or} \quad (.9811, 1.1853).$$

With 95% confidence we can conclude that for each unit increase in initial oyster weight there is an increase in final weight of between .9811 and 1.1853.

Chapter 13

Completely Randomized Factorial Experiments

13.1. (a) Each of the two contrasts can be estimated with a linear combination of treatment means. Then the standard error of the contrasts can be estimated. An α-level is chosen and a critical value is determined as $t(\text{df}, \alpha/k)$, since they are one-sided tests. Each test statistic is compared to the critical value and a conclusion is drawn.

(b) From Example 9.14 the point estimates are

$$\hat{\theta}_1 = \overline{Y}_{FW} - \overline{Y}_{SW} = 2.01 - 1.59 = 0.42$$

$$\hat{\theta}_2 = \overline{Y}_{FC} - \overline{Y}_{SC} = 1.08 - 1.55 = -0.47$$

Using an $\alpha = 0.05$ the t-statistic with eight degrees of freedom is $t(8, 0.05/2) = t(8, 0.025) = 2.306$. Because we have equal replications the standard errors are

$$\hat{\sigma}_{\hat{\theta}_1} = \hat{\sigma}_{\hat{\theta}_2} = \sqrt{\left(\frac{1^2}{3} + \frac{(-1)^2}{3}\right)0.034} = 0.15055.$$

The test statistics are

$$t_1 = \frac{\hat{\theta}_1}{\hat{\sigma}_{\hat{\theta}_1}} = \frac{0.42}{0.15055} = 2.7898$$

and

$$t_2 = \frac{\hat{\theta}_2}{\hat{\sigma}_{\hat{\theta}_2}} = \frac{-0.47}{0.15055} = -3.1219$$

Since $t_1 > 2.306$ and $t_2 < -2.306$ we conclude that at the 0.05 level there is sufficient evidence to reject the hypotheses that

$$\mu_{FW} - \mu_{SW} = 0, \quad \mu_{FC} - \mu_{SC} = 0$$

and conclude that

$$\mu_{FW} - \mu_{SW} > 0, \quad \text{and} \quad \mu_{FC} - \mu_{SC} < 0.$$

(c) The Bonferroni simultaneous confidence interval method will be done. This method will use $t(8, 0.05/4) = 2.7515$. Hence, the intervals are

$$\hat{\theta}_1 \pm t(8, 0.05/4)\hat{\sigma}_{\hat{\theta}_1} = 0.42 \pm 2.7515(0.15055) = (0.0058, 0.8342)$$

and

$$\hat{\theta}_2 \pm t(8, 0.05/4)\hat{\sigma}_{\hat{\theta}_2} = -0.47 \pm 2.7515(0.15055) = (-0.8842, -0.0558).$$

With 95% confidence we conclude that the true difference in mean gain in weight of fish grown in flowing water versus still water at the warm temperature is between 0.0058 and 0.8342 and that the true difference in mean gain in weight of fish grown in flowing water versus still water at the cold temperature is between -0.8842 and -0.0558.

(d) $\theta = (\mu_{FW} - \mu_{SW}) - (\mu_{FC} - \mu_{SC})$

(e) The point estimate is

$$\hat{\theta} = (\overline{Y}_{FW} - \overline{Y}_{SW}) - (\overline{Y}_{FC} - \overline{Y}_{SC}) = 0.42 - (-0.47) = 0.89$$

and the standard error is

$$\hat{\sigma}_{\hat{\theta}} = \sqrt{\left(\frac{1^2}{3} + \frac{(-1)^2}{3} + \frac{(-1)^2}{3} + \frac{1^2}{3}\right) 0.034} = 0.2129.$$

We then have

H_0: $\theta = 0$

H_1: $\theta \neq 0$

Test Statistic: $t_c = \dfrac{\hat{\theta}}{\hat{\sigma}_{\hat{\theta}}} = \dfrac{0.89}{0.2129} = 4.1804$

Rejection Region: Reject H_0 if $|t_c| > t(8, 0.025) = 2.306$

Conclusion: Reject H_0. The data shows at the 5% significance level that the differential gain in weight resulting from still and flowing waters are not the same at the two temperatures.

13.3. (a) The two factors of interest are Age and Level of Processing. The levels for Age are Young and Old. The levels for Level are Counting, Rhyming, Adjective, Imagery, and Intentional.

(b) A = Age a_1 = young, a_2 = old
B = Level of Processing b_1 = counting, b_2 = rhyming, b_3 = adjective, b_4 = Imagery, b_5 = Intentional.
The study is a 2 × 5 complete factorial with $t = 10$ treatments. The treatment $a_1 b_3$ corresponds to a young subject processing at the adjective level.

(c) Two research hypotheses that can be verified using a factorial analyses are

   i. The difference in the young and old subjects' ability to recall words depend on the level of processing.
   ii. The difference in the subject's ability depends on their age.

13.5. (a) There are three factors involved in the study; PRETREATMENT Method with levels 1, 2 and 3, CUTTING Number with levels 1, 2 and 3, and VARIETY of Sweet Potato, with levels(types) 1, 2, 3 and 4.

(b) A useful set of notation would be :
A=PRETREATMENT Method, a_1=Pretreatment 1, a_2=Pretreatment 2, a_3=Pretreatment 3.
B=CUTTING Number, b_1=Cutting 1, b_2=Cutting 2, b_3=Cutting 3.
C=VARIETY, c_1=Variety 1, ..., c_4=Variety 4.

(c) One hypothesis that could be tested is "Does the difference between Pretreatments 1 and 2 change for different varieties?" Another hypothesis is "Does Variety 1 differ from Variety 2 when averaged over all Cuttings and all Varieties?"
There are many different hypotheses that could be tested. However, the specific hypotheses should be decided on before the data are collected or looked at.

13.7. The three plots take the following form.

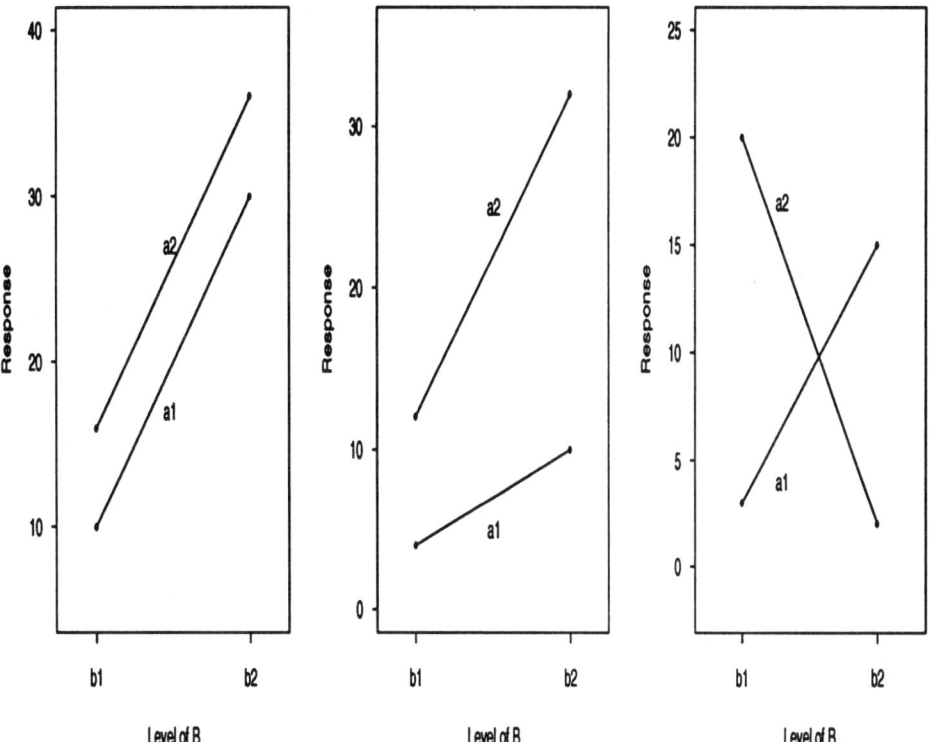

The plots lead to the exact same conclusions as those in the text. This is the case because the interpretation of interaction does not depend on one factor alone, but on the combinations of the two factors. The plots can be thought of as just a change in the axis between the two factors.

13.9. (a) Using the calculations from the previous chapters the ANOVA table has the following form.

Source	df	SS	MS	F_c
Treatment	3	1086.00	362.00	37.3581
Error	33	319.77	9.69	
Total	36	1405.77		

F_c is then compared to an $F(3, 33)$ and has associated $p-$value$< .0001$. Since the associated $p-$value is smaller than most reasonable α–levels, it is concluded that the four groups differ with respect to mean reduction in their cholesterol levels.

(b) A contrast that measures the effect of interaction between age and gender has the following form.

$$\mu_{[AB]} = \frac{1}{2}\left(-(\mu_{OM} - \mu_{OF}) + (\mu_{YM} - \mu_{YF})\right)$$
$$= \frac{1}{2}(\mu_1 - \mu_2 - \mu_3 + \mu_4).$$

The estimated value of the contrast is calculated by replacing the population means by their sample counterparts.

141

(c)

$$\hat{\mu}_{[AB]} = \frac{1}{2}\left((12-14)-(26-21)\right) = \frac{1}{2}[-7] = -3.5,$$

$$SS[AB] = \frac{(-3.5)^2}{\left[\frac{(\frac{-1}{2})^2}{8} + \frac{(\frac{1}{2})^2}{10} + \frac{(\frac{1}{2})^2}{9} + \frac{(\frac{-1}{2})^2}{10}\right]} = \frac{12.25}{.1090} = 112.3567,$$

$$F_{cal} = \frac{MS[AB]}{MS[E]} = \frac{112.3567/1}{9.69} = 11.5951.$$

F_{cal} is compared to an $F(1,33)$ and has associated p-value .0017. At the 0.01 level, there is evidence to conclude that there is interaction between age and gender. The effect of age seems to depend on which gender the person is in terms of the amount of reduction in cholesterol level.

(d) A 95% confidence interval for the interaction has the following form

$$\hat{\mu}_{[AB]} \pm t(33,.025)\sqrt{MS[E]\left(\frac{(\frac{-1}{2})^2}{8} + \frac{(\frac{1}{2})^2}{10} + \frac{(\frac{1}{2})^2}{9} + \frac{(\frac{-1}{2})^2}{10}\right)}.$$

Upon substitution this becomes

$$-3.5 \pm 2.0345\sqrt{9.69(.1090)} \quad \text{or} \quad (-5.5909, -1.4091).$$

With 95% confidence we conclude that the true interaction effect is in the interval (-5.5909, -1.4091).

13.11. (a) The appropriate contrasts can be written in terms of the population means. The estimated contrasts are obtained by replacing the population means with their sample means.

$$\mu_{[AGE]} = \frac{1}{2}\left[(\mu_{OM} - \mu_{YM}) + (\mu_{OF} - \mu_{YF})\right]$$

$$= \frac{1}{2}\left[(\mu_3 - \mu_1) + (\mu_4 - \mu_2)\right]$$

$$\mu_{[GENDER]} = \frac{1}{2}\left[(\mu_{OM} - \mu_{OF}) + (\mu_{YM} - \mu_{YF})\right]$$

$$= \frac{1}{2}\left[(\mu_3 - \mu_4) + (\mu_1 - \mu_2)\right]$$

(b) Previous calculations from Exercise 13.9 give us $MS[E] = 9.69$ with 33 df. Also it is easy to verify that the denominator in the $SS[AGE]$ and $SS[GENDER]$ are the same as the denominator of $SS[AB]$ calculated previously.

$$\hat{\mu}_{[AGE]} = \frac{1}{2}\left[(12-26) + (14-21)\right] = \frac{-21}{2} = -10.5$$

$$\hat{\mu}_{[GENDER]} = \frac{1}{2}\left[(12-14) + (26-21)\right] = \frac{3}{2} = 1.5$$

$$SS[AGE] = \frac{(-10.5)^2}{0.1090} = 1011.4679$$

$$SS[GENDER] = \frac{1.5^2}{0.1090} = 20.6422$$

The test for the main effect of AGE becomes

$$F_{cal} = \frac{MS[AGE]}{MS[E]} = \frac{1011.4679/1}{9.69} = 104.3827.$$

142

This is compared to a $F(1, 33)$ and has associated p–value < 0.0001. The test for the main effect of GENDER becomes

$$F_{cal} = \frac{MS[GENDER]}{MS[E]} = \frac{20.6422/1}{9.69} = 2.1303.$$

This is compared to an $F(1, 33)$ and has associated p–value $= 0.1539$. No real conclusions should be made on the main effects since in Exercise **13.9** the interaction effect was found to be significant.

(c) Simultaneous confidence intervals for the three effects are constructed using the Bonferroni approach. The contrasts have the following form

$$\widehat{\mu}_{\text{effect}} \pm t(33, 0.025/2(3)) \sqrt{MS[E](0.1090)}.$$

For the interaction effect we get

$$-3.5 \pm 2.7174(1.0278) \quad \text{or} \quad (-6.2929, -0.7071).$$

For the main effect of AGE we get

$$-10.5 \pm 2.7174(1.0278) \quad \text{or} \quad (-13.2929, -7.7071).$$

For the main effect of GENDER we get

$$1.5 \pm 2.7174(1.0278) \quad \text{or} \quad (-1.2929, 4.2929).$$

With 95% confidence we conclude that simultaneously each of the true effects $\mu_{[AB]}$, $\mu_{[AGE]}$ and $\mu_{[GENDER]}$ is within its respective interval.

13.13. (a) As in the text, we will disregard the data for 100% humidity for this exercise. Denote Temperature by A and Humidity by B. Then, the AB interaction effect, written as the difference between the simple effects of A, is

$$\mu[AB] = \frac{1}{2}\left(\mu[AB_2] - \mu[AB_1]\right).$$

These simple effects are

$$\mu[AB_1] = \mu_{21} - \mu_{11} \quad \mu[AB_2] = \mu_{22} - \mu_{12}$$

which are estimated as

$$\hat{\mu}[AB_1] = \overline{Y}_{21+} - \overline{Y}_{11+} = 69.50 - 72.50 = -3.00 \quad \hat{\mu}[AB_2] = \overline{Y}_{22+} - \overline{Y}_{12+} = 78.25 - 81.50 = -3.25.$$

Then, the estimate of the interaction effect is

$$\hat{\mu}[AB] = \frac{1}{2}\left(-3.25 - (-3.00)\right) = -0.25.$$

With this information, the sum of squares is

$$SS[AB] = n\left(\hat{\mu}[AB]\right)^2 = 4(-0.25)^2 = 0.25.$$

Comparing this to the value for MS[E] given in the text, 33.31, the F-statistic is

$$F = \frac{SS[AB]/df}{MS[E]} = \frac{0.25/1}{33.31} = 0.0075.$$

Using Table C.4 this value is much smaller than any tabled value. Therefore, its p–value is much larger than 0.100. We can not conclude that there is an effect of interaction between humidity and temperature.

143

(b) To simultaneously test the three effects (interaction and both main effects) with a probability of a Type I error below 1%, we can compare each p-value to the alpha level found via the Bonferroni approach. In other words we would not compare each p-value to .01 but we would compare each to $\frac{.01}{3} = 0.003$. The three tests have associated p-values of 0.9669, 0.0096 and 0.3007 for interaction, Temperature and Humidity respectively. Since each of these p-values is > 0.003, when all three effects are tested simultaneously, there is insufficient evidence at the 0.01 level that any of the three effects have any significant effect on Percent Water Content of Snail Tissue.

13.15. (a) The estimating equations for the three contrasts can be found in the text and are repeated here.

$$\hat{\mu}_{[A]} = \frac{1}{2}\overline{Y}_{22+} + \frac{1}{2}\overline{Y}_{21+} - \frac{1}{2}\overline{Y}_{12+} - \frac{1}{2}\overline{Y}_{11+}$$

$$\hat{\mu}_{[B]} = \frac{1}{2}\overline{Y}_{22+} - \frac{1}{2}\overline{Y}_{21+} + \frac{1}{2}\overline{Y}_{12+} - \frac{1}{2}\overline{Y}_{11+}$$

$$\hat{\mu}_{[AB]} = \frac{1}{2}\overline{Y}_{22+} - \frac{1}{2}\overline{Y}_{21+} - \frac{1}{2}\overline{Y}_{12+} + \frac{1}{2}\overline{Y}_{11+}$$

Using the definition of orthogonal contrasts, we find that the contrasts $\hat{\mu}_{[A]}$ and $\hat{\mu}_{[B]}$ are orthogonal if

$$\frac{\frac{1}{2}\left(\frac{1}{2}\right)}{n_{22}} + \frac{\frac{1}{2}\left(-\frac{1}{2}\right)}{n_{21}} + \frac{-\frac{1}{2}\left(\frac{1}{2}\right)}{n_{12}} + \frac{-\frac{1}{2}\left(-\frac{1}{2}\right)}{n_{11}} = 0.$$

If all $n_{ij} = n$ then the left hand side of the above equation becomes

$$\frac{1}{n}\left[\frac{1}{2}\left(\frac{1}{2}\right) + \frac{1}{2}\left(-\frac{1}{2}\right) + \left(-\frac{1}{2}\right)\left(\frac{1}{2}\right) + \left(\frac{1}{2}\right)\left(-\frac{1}{2}\right)\right] = \frac{1}{n}\left[\frac{1}{4} - \frac{1}{4} - \frac{1}{4} + \frac{1}{4}\right] == 0.$$

Hence the two contracts are orthogonal. Similar calculation for the pairs $\hat{\mu}_{[B]}$ with $\hat{\mu}_{[AB]}$ and $\hat{\mu}_{[A]}$ with $\hat{\mu}_{[AB]}$ lead to the same conclusion. Thus since all pairs of contrasts are orthogonal, the set of contrasts is said to be mutually orthogonal.

(b) For testing the orthogonality of $\hat{\mu}_{[A]}$ with $\hat{\mu}_{[B]}$ we get

$$\frac{\frac{1}{4}}{3} - \frac{\frac{1}{4}}{3} - \frac{\frac{1}{4}}{3} + \frac{\frac{1}{4}}{2} = -\frac{1}{12} + \frac{1}{8} = \frac{1}{24} \neq 0.$$

Hence the pair $\hat{\mu}_{[A]}$ and $\hat{\mu}_{[B]}$ are not orthogonal. Similar calculations lead to the same conclusions for the other two pairs. Thus, the set is not mutually orthogonal.

13.17. (a) Using the notation of the equations in the text with n_i replacing n we get

$$
\begin{aligned}
SS[Trt] &= \frac{Y_{YM}^2}{n_1} + \frac{Y_{YF}^2}{n_2} + \frac{Y_{OM}^2}{n_3} + \frac{Y_{OF}^2}{n_4} - \frac{Y_{++}^2}{n_1 + n_2 + n_3 + n_4} \\
&= \frac{(8(26))^2}{8} + \frac{(10(21))^2}{10} + \frac{(9(12))^2}{9} + \frac{(10(14))^2}{10} - \\
&\qquad \frac{(8(26) + 10(21) + 9(12) + 10(14))^2}{37} = 13074 - 11988 = 1086 \\
SS[A] &= \frac{Y_Y^2}{n_1 + n_2} + \frac{Y_O^2}{n_3 + n_4} - \frac{Y_{++}^2}{n_1 + n_2 + n_3 + n_4} \\
&= \frac{(8(26) + 10(21))^2}{18} + \frac{(9(12) + 10(14))^2}{19} - 11988 \\
&= 12943.94 - 11988 = 955.9415 \\
SS[G] &= \frac{Y_M^2}{n_1 + n_3} + \frac{Y_F^2}{n_2 + n_4} - \frac{Y_{++}^2}{n_1 + n_2 + n_3 + n_4}
\end{aligned}
$$

$$
\begin{aligned}
&= \frac{(8(26)+9(12))^2}{17} + \frac{(10(21)+10(14))^2}{20} - 11988 \\
&= 11998.88 - 11988 = 10.8823 \\
SS[Inter] &= SS[Trt] - SS[A] - SS[G] \\
&= 1086 - 955.9415 - 10.8823 = 119.1762
\end{aligned}
$$

The sums of squares calculated here are different (and incorrect) than those previously calculated. The reason for this is that when the sample sizes are unequal the sums of squares cannot be calculated using Equation 13.12.

(b) Using the calculations made in the previous Exercises we get the following ANOVA table.

Source	df	SS	MS	F_c
AGE	1	1011.47	1011.47	104.38
GENDER	1	20.64	20.64	2.13
Inter	1	112.36	112.36	11.59
Error	33	319.77	9.69	
Total	36	1405.77		

(c) Based on the ANOVA table and previous Exercises, we can conclude that there is strong evidence that there is interaction between AGE and GENDER in determining the drop in cholesterol level. The associated p−value was found in Exercise **13.9** to be .0017. The main effects should not be analyzed since the interaction is significant.

13.19. (a) For the data set having No Interaction, the components of the simple effects of B are:

Components of the simple effect	Level of B(j)		
	b_1	b_2	b_3
$\mu_1[A_jB]$	−10	−10	−10
$\mu_2[A_jB]$	−20	−20	−20

For the data set having Quantitative interaction, the components of the simple effects of B are:

Components of the simple effect	Level of B(j)		
	b_1	b_2	b_3
$\mu_1[A_jB]$	−10	−10	−10
$\mu_2[A_jB]$	−20	−30	−40

For the data set having Qualitative interaction, the components of the simple effects of B are:

Components of the simple effect	Level of B(j)		
	b_1	b_2	b_3
$\mu_1[A_jB]$	−10	−10	−10
$\mu_2[A_jB]$	−20	−80	30

(b) As can be seen in the above tables, when there is no interaction present, the simple effects for both A and B are identical within rows. When there is quantitative interaction between the effects, the simple effects for both A and B are not constant within rows, but have the same sign. When there is qualitative interaction between the effects, the simple effects for both A and B are not identical within rows and some simple effects have different signs.

145

13.21. (a) The three simple effects of water movement are found as follows:

$$\hat{\mu}[A_1B] = \overline{Y}_{12} - \overline{Y}_{11} = \frac{3.24}{3} - \frac{4.65}{3} = -0.47$$

$$\hat{\mu}[A_2B] = \overline{Y}_{22} - \overline{Y}_{21} = \frac{5.28}{3} - \frac{5.61}{3} = -0.11$$

$$\hat{\mu}[A_3B] = \overline{Y}_{32} - \overline{Y}_{31} = \frac{6.03}{3} - \frac{4.77}{3} = 0.42$$

All of the standard errors for the simple effects are the same. They take the value

$$\hat{\sigma}_{\hat{\mu}[A_iB]} = \sqrt{MS[E]\left(\frac{1}{3} + \frac{1}{3}\right)} = \sqrt{0.025\left(\frac{2}{3}\right)} = 0.1291.$$

Using the Bonferroni approach for simultaneous confidence intervals, the intervals all have the same form

$$\hat{\mu}[A_iB] \pm t(12, \tfrac{0.10}{6})\hat{\sigma}_{\hat{\mu}[A_iB]}.$$

For Temperature 1(cold) we get

$$-0.47 \pm 2.4029(0.1291) \quad \text{or} \quad (-0.7802, -0.1598).$$

For Temperature 2(lukewarm) we get

$$-0.11 \pm 2.4029(0.1291) \quad \text{or} \quad (-0.4202, 0.2002).$$

For Temperature 3(warm) we get

$$0.42 \pm 2.4029(0.1291) \quad \text{or} \quad (0.1098, 0.7302).$$

(b) Based on the above confidence intervals, we can argue that with 90% confidence we can conclude that simultaneously there is an effect of water movement for both the cold and the warm temperatures (these intervals do not contain zero), but in opposite directions, and no effect of water movement for the lukewarm temperature. To use the p-values in the text for the comparisons, these p-values should be compared to the α-level $\frac{.10}{6} = .01667$. Only the p-values for the cold and warm temperatures fall below the alpha level, so the identical conclusions can be made based on the tests in the text.

(c) To apply the Tukey multiple comparison method for the three mean weight gains at each of the two water movement levels we will compare each pair of differences to the critical value, using $\alpha = 0.05$,

$$q(t, N-t, \alpha)\sqrt{MS[E]\left(\frac{1}{3}\right)} = q(3, 12, 0.05)\sqrt{\frac{0.025}{3}} = 3.77(0.0913) = 0.3442.$$

The following table can be used to compare the simple effects.

Comparison	Still Water		Flowing Water	
	Abs. Diff.	Sign.	Abs. Diff.	Sign.
Temp1 − Temp2	$\frac{\|4.65-5.61\|}{3} = 0.32$	NS	$\frac{\|3.24-5.28\|}{3} = 0.68$	S
Temp2 − Temp3	$\frac{\|5.61-4.77\|}{3} = 0.28$	NS	$\frac{\|5.28-6.03\|}{3} = 0.25$	NS
Temp1 − Temp3	$\frac{\|4.65-4.77\|}{3} = 0.04$	NS	$\frac{\|3.24-6.03\|}{3} = 0.93$	S

The table can be summarized as

$$\text{Still} \quad \underline{1, 2, 3} \qquad \text{Flowing} \quad 1, \underline{2, 3}$$

This table shows how interaction seems present. The differences of the means for Temperature seem to depend on Water Movement.

13.23. Before beginning this Exercise we construct a table of treatment totals for the twelve treatment combinations.

Pretreatment	Variety Type				
Level	1	2	3	4	Total
1	23	54	68	18	163
2	36	61	72	37	206
3	10	13	13	25	61
Total	69	128	153	80	430

(*) Totals are based on n=3 replications per treatment

(a) Using the calculations from the previous chapters the ANOVA table has the following form.

Source	df	SS	MS	F_c
Treatment	11	1872.5556	170.2323	8.5235
Error	24	479.3333	19.9722	
Total	35	2351.8889		

F_c is then compared to an $F(11, 24)$ and has associated p–value$< .0001$. Since the associated p–value is smaller than all reasonable α–levels, it is concluded that the twelve treatment combinations have different expected root lengths.

(b) The following plot can help in looking for interaction and main effects.

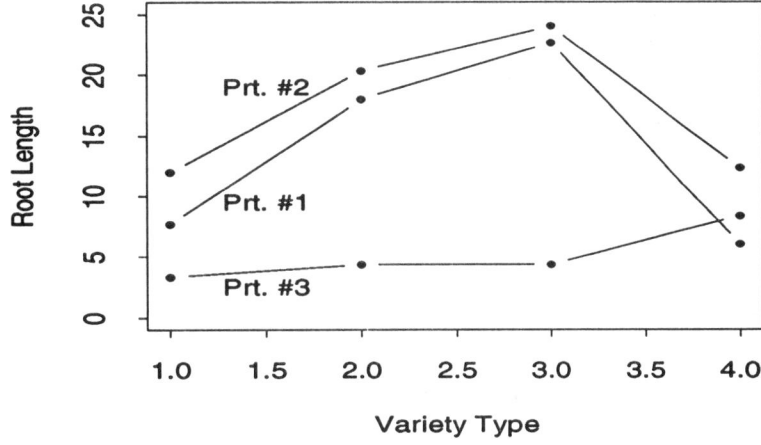

Based on the plot there seems to be some evidence of possible interaction between variety and pretreatment type.

(c) The following ANOVA table was found using SAS and PROC ANOVA.

147

Dependent Variable: LENGTH

Source	DF	Sum of Squares	Mean Square	F Value	Pr > F
Model	11	1872.5556	170.2323	8.52	0.0001
Error	24	479.3333	19.9722		
Corrected Total	35	2351.8889			

R-Square	C.V.	Root MSE	LENGTH Mean
0.796192	37.41513	4.4690	11.944

Source	DF	Anova SS	Mean Square	F Value	Pr > F
PRETRT	2	924.38889	462.19444	23.14	0.0001
VARIETY	3	525.44444	175.14815	8.77	0.0004
PRETRT*VARIETY	6	422.72222	70.45370	3.53	0.0120

(d) The first thing that needs to be tested in a factorial analysis is the interaction effect. The interaction effect is found to have a p–value of .0120. If an α–level of .05 is used then the interaction term is found to be significant. If an α–level of .01 is used then the interaction term is found not to be significant. The decision as to whether to look at main effects or simple effects hinges on this decision. Since the p–value is close to .01 simple effect analysis would most likely be the safest and most appropriate thing to do. The twelve simple effects would be tested in a manner similar to that done in Exercise **13.21**.

13.25. This model can be written as

$$Y_{ijk} = \mu + \alpha_i + \beta_j + (\alpha\beta)_{ij} + E_{ijk} \qquad i = 1, 2; j = 1, 2; k = 1, \ldots, n_{ij}$$

where

Y_{ijk} = the percent reduction in cholesterol level after 6 months on the experimental diet for the ith GENDER (i=1–Male, i=2–Female), j^{th} AGE (j=1–Young, j=2–Old), and k^{th} SUBJECT (k=1,...,n_{ij})

μ = overall mean percent reduction in cholesterol level,

α_i = the effect of the ith level of SEX; $\alpha_1 + \alpha_2 = 0$

β_j = the effect of the jth level of AGE; $\beta_1 + \beta_2 = 0$

$(\alpha\beta)_{ij}$ = the joint effect of the ith level of SEX and the jth level of AGE; $(\alpha\beta)_{1j} + (\alpha\beta)_{2j} = 0, j = 1, 2$; and $(\alpha\beta)_{i1} + (\alpha\beta)_{i2} = 0, i = 1, 2$

E_{ijk} = the random error. The E_{ijk} are independent normal with mean 0 and variance σ^2.

13.27. Let Y_{ijk} denote the percent water content in tissue of snails by the ith TEMPERATURE (i=1–20°C, i=2–30°C), j^{th} level of HUMIDITY (j=1–45%, j=2–75%, j=3–100%), and k^{th} REPLICATE (k=1,...,4). Then

$$Y_{ijk} = \mu + \alpha_i + \beta_j + (\alpha\beta)_{ij} + E_{ijk} \quad i = 1, 2; j = 1, 2, 3; k = 1, \ldots, 4,$$

where

μ = overall mean percent water content in tissue

$$\alpha_i \quad = \quad \text{the effect of the } i^{th} \text{ TEMPERATURE;} \sum_i \alpha_i = 0,$$

$$\beta_j \quad = \quad \text{the effect of the } j^{th} \text{ level of HUMIDITY;} \sum_j \beta_j = 0,$$

$$(\alpha\beta)_{ij} \quad = \quad \text{the joint effect of the } i^{th} \text{ TEMPERATURE and the } j^{th} \text{ level of HUMIDITY;}$$
$$\sum_i (\alpha\beta)_{ij} = \sum_j (\alpha\beta)_{ij} = 0,$$

$$E_{ijk} \quad = \quad \text{the random error. The } E_{ijk} \text{ are independent normal with mean 0 and variance } \sigma^2.$$

13.29. (a) The following two plots give information about interaction between factors A and C for each level of B.

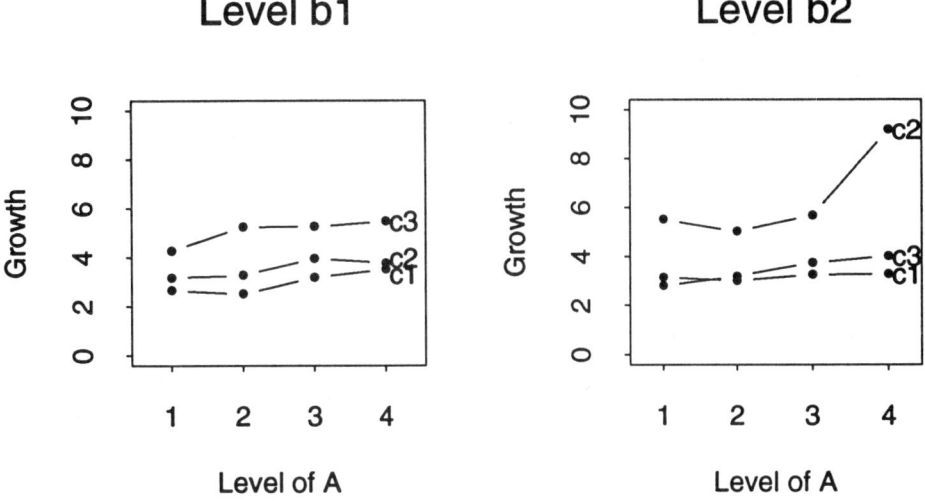

Using the notation in the text, the interaction can be explained as ACB_1 absent, ACB_2 present and ABC present.

(b) Let Y_{ijkl} denote the amount of fungus growth by the ith TIME PERIOD (i=1,...,4), j^{th} ENVIRONMENTAL FACTOR (j=1, 2), k^{th} GROWTH MEDIA (k=1,/ldots,3), and l^{th} REPLICATE(l=1, 2). Then

$$Y_{ijkl} \quad = \quad \mu + \alpha_i + \beta_j + \gamma_k + (\alpha\beta)_{ij} + (\alpha\gamma)_{ik} + (\beta\gamma)_{jk} + (\alpha\beta\gamma)_{ijk} + E_{ijkl}$$
$$i = 1, 2; j = 1, 23; k = 1, \ldots, 4,$$

where

$$\mu \quad = \quad \text{overall mean amount of fungus growth}$$
$$\alpha_i \quad = \quad \text{the effect of the } i^{th} \text{ TIME PERIOD}$$
$$\beta_j \quad = \quad \text{the effect of the } j^{th} \text{ ENVIRONMENTAL FACTOR}$$
$$\gamma_k \quad = \quad \text{the effect of the } k^{th} \text{ GROWTH MEDIA}$$
$$(\alpha\beta)_{ij} \quad = \quad \text{the joint effect of the } i^{th} \text{ TIME PERIOD and the } j^{th}$$
$$\text{ENVIRONMENTAL FACTOR}$$
$$(\alpha\gamma)_{ik} \quad = \quad \text{the joint effect of the } i^{th} \text{ TIME PERIOD and the } k^{th}$$
$$\text{GROWTH MEDIA}$$
$$(\beta\gamma)_{jk} \quad = \quad \text{the joint effect of the } j^{th} \text{ ENVIRONMENTAL FACTOR and the } k^{th}$$
$$\text{GROWTH MEDIA}$$

$(\alpha\beta\gamma)_{ijk}$ = the joint effect of the ith TIME PERIOD, j^{th} ENVIRONMENTAL FACTOR and the kth GROWTH MEDIA

E_{ijkl} = the random error.

(c) Confidence intervals for the difference between effects of the two environments can be accomplished by confidence intervals on the difference between the two means. The six necessary means (each with 8 replicates) and their differences are in the following table.

Growth Medium(C)	Environment(B)		
	b_1	b_2	diff($b_2 - b_1$)
c_1	2.9575	3.1663	0.2088
c_2	3.5375	6.33	2.7925
c_3	5.0513	3.4313	-1.62

To complete the confidence intervals we need estimates of the standard error of each mean difference. The estimated standard error for the mean difference is

$$\hat{\sigma} = \sqrt{MS[E]\left(\frac{1}{8} + \frac{1}{8}\right)} = \sqrt{2.06(0.25)} = 0.7176.$$

Hence, each confidence interval has the following form

$$\overline{Y}_{+2k} - \overline{Y}_{+1k} \pm t(24, 0.025)(0.7176).$$

For Growth Medium c_1 this becomes

$$0.2088 \pm 2.064(0.7176) \quad \text{or} \quad (-1.2723, 1.6899).$$

For Growth Medium c_2 this becomes

$$2.7925 \pm 2.064(0.7176) \quad \text{or} \quad (1.3114, 4.2736).$$

For Growth Medium c_3 this becomes

$$-1.62 \pm 2.064(0.7176) \quad \text{or} \quad (-3.1011, -0.1389).$$

With 95% confidence we conclude that for Growth Medium c_1 there is no difference between the two Environments. With 95% confidence we conclude that for Growth Medium c_2 Environment b_2 increases fungus growth between 1.314 and 4.2736 more than Environment b_3. With 95% confidence we conclude that for Growth Medium c_3 Environment b_1 increases fungus growth between 3.1011 and 0.1389 more than Environment b_2.

(d) The multiple comparisons will be made using the Tukey method. Within each Environmental factor the absolute mean differences will be compared to the value

$$q(t, dferror, \alpha)\sqrt{\frac{MS[E]}{n}} = q(3, 24, .05)\sqrt{\frac{2.06}{8}} = 3.53(0.5074) = 1.7913.$$

The individual comparisons can be made using the following table.

Environment	c_3 vs. c_1	c_3 vs. c_2	c_2 vs. c_1
b_1	2.0938 (S)	1.5138 (NS)	0.58 (NS)
b_2	0.2650 (NS)	-2.8987 (S)	3.1637 (S)

This once again informs us of interaction between the two effects because differences between Growth Mediums depend on which Environment is involved.

13.31. (a) The interaction plot for the second order interaction is

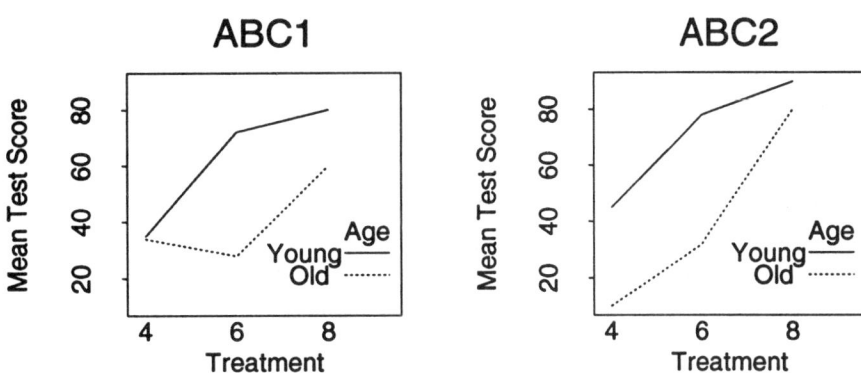

The interaction plot for the first order interactions is

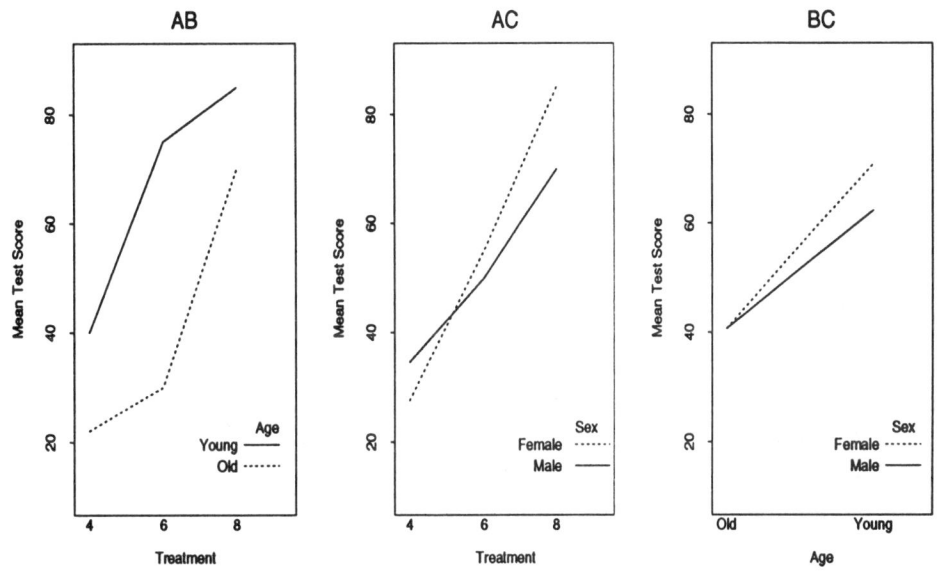

(b) The plot for Males (ABC1) shows two non-parallel lines for Young and Old males. This indicates that there is an interaction between Age and Treatment for this level of Gender. If we see the exact same relationship for Females, then we would only conclude the first order and not the second order interaction exists.

However, the lines representing the mean Test Scores of Young and Old females at each level of the Treatment do not exhibit the same behavior. The mean Test Scores for the Female subjects allowed four hours of sleep are farther apart than the mean Test Scores for the Male subjects allowed four hours of sleep. The different behavior indicates that there might be a second order interaction.

Each of the three plots of possible first order interaction show lines that are not parallel. This indicates that each might have a significant first order interaction.

(c) **H₀:** No Age×Gender×Treatment interaction exists.

H₁: There is an Age×Gender×Treatment interaction.

151

Test Statistic: $F_c = \dfrac{388}{143} = 2.7133$

Rejection Region: Reject H_0 if $F_c > F(2, 24, 0.05) = 3.4028$

Conclusion: Do not reject H_0. At the 0.05 significance level, there is not sufficient evidence to indicate that an Age×Gender×Treatment interaction exists.

H_0: No Gender×Treatment interaction exists.

H_1: There is a Gender×Treatment interaction.

Test Statistic: $F_c = \dfrac{364}{143} = 2.5454$

Rejection Region: Reject H_0 if $F_c > F(2, 24, 0.05) = 3.4028$

Conclusion: Do not reject H_0. At the 0.05 significance level, there is not sufficient evidence to indicate that a Gender×Treatment interaction exists.

H_0: No Age×Treatment interaction exists.

H_1: There is an Age×Treatment interaction.

Test Statistic: $F_c = \dfrac{819}{143} = 5.7272$

Rejection Region: Reject H_0 if $F_c > F(2, 24, 0.05) = 3.4028$

Conclusion: Reject H_0. At the 0.05 significance level, there is sufficient evidence to indicate that a Age×Treatment interaction exists.

H_0: No Age×Gender interaction exists.

H_1: There is an Age×Gender interaction.

Test Statistic: $F_c = \dfrac{169}{143} = 1.1818$

Rejection Region: Reject H_0 if $F_c > F(1, 24, 0.05) = 4.2597$

Conclusion: Do not reject H_0. At the 0.05 significance level, there is not sufficient evidence to indicate that a Age×Gender interaction exists.

H_0: No Gender effect exists.

H_1: A Gender effect exists.

Test Statistic: $F_c = \dfrac{169}{143} = 1.1818$

Rejection Region: Reject H_0 if $F_c > F(1, 24, 0.05) = 4.2597$

Conclusion: Do not reject H_0. At the 0.05 significance level, there is not sufficient evidence to indicate that a Gender effect exists.

These are all of the tests that are appropriate.

(d) The analysis does not contradict the conclusions in Part b. A second order interaction looked possible, but more important is the strength of the interaction compared to the mean square error. The plots gives us important indications and ideas, but the test is the true measure.

(e) The Tukey multiple pairwise comparison method of Chapter 9 will be used. (We can use this because the sample sizes are equal.) First, the mean test scores of the three treatments will be compared for each age group. Second, the mean test scores of the two age groups will be compared for each of the three treatments. This approach is taken because it is not meaningful to compare all pairs of means, for example, Young:4hrs with Old:6hrs.

For the first comparisons, there are $t = 3$ treatments involved for each age level. The treatment means are

	4	6	8
Y	40	75	85
O	22	30	70

152

The LSD value is

$$LSD_{ij}(T) = q(3, 24, 0.05)\sqrt{\frac{MS[E]}{n}} = 3.53\sqrt{\frac{143}{6}} = 17.2332.$$

Using this value the comparisons show

$$\text{Young:} \quad \text{4hrs} \quad \overline{\text{6hrs} \quad \text{8hrs}}$$

and

$$\text{Old:} \quad \overline{\text{4hrs} \quad \text{6hrs}} \quad \text{8hrs}$$

(f) Yes, the multiple comparison is justified. The analysis showed a statistically significant Age×Gender interaction.

13.33. (a) The model can be written as

$$Y_{ijk} = \mu + \alpha_i + \beta_j + (\alpha\beta)_{ij} + E_{ijk} \qquad i = 1, 2; j = 1, 2, 3; k = 1, \dots, n_{ij}$$

where

Y_{ijk} = the measured yield for the kth plant given the ith method of Fertilization and the jth method of Irrigation,

μ = overall mean yield,

α_i = fixed effect of ith method of Fertilization,

β_j = fixed effect of jth method of Irrigation,

$(\alpha\beta)_{ij}$ = joint effect of the ith method of Fertilization and the jth method of Irrigation,

E_{ijk} = random error distributed as independent $N(0, \sigma^2)$.

(b) The reduced models associated with each of the null hypotheses are

i. $Y_{ijk} = \mu + \beta_j + (\alpha\beta)_{ij} + E_{ijk}$

ii. $Y_{ijk} = \mu + \alpha_i + (\alpha\beta)_{ij} + E_{ijk}$

iii. $Y_{ijk} = \mu + \alpha_i + \beta_j + E_{ijk}$

(c) The Type III sum of squares labeled FERT is the difference between the regression sum of squares associated with the full model and the regression sum of squares associated with the reduced model *i* in part b.

The Type III sum of squares labeled IRRIG is the difference between the regression sum of squares associated with the full model and the regression sum of squares associated with the reduced model *ii* in part b.

The Type III sum of squares labeled FERT×IRRIG is the difference between the regression sum of squares associated with the full model and the regression sum of squares associated with the reduced model *iii* in part b.

13.35. (a) The tree diagram has the following form.

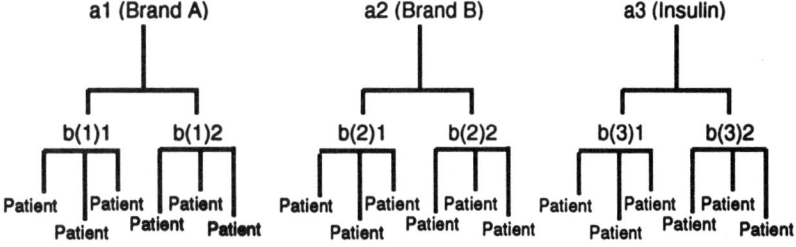

(b) The nested ANOVA model can be written as

$$Y_{ijk} = \mu + \alpha_i + \beta_{(i)j} + E_{ijk} \qquad i = 1, 2, 3; j = 1, 2; k = 1, 2, 3,$$

where

$$Y_{ijk} = \text{the change in the blood sugar for the } i^{th} \text{ Brand and Injection,}$$
$$j^{th} \text{ Administration Method for the } k^{th} \text{ Patient,}$$
$$\mu = \text{overall mean,}$$
$$\alpha_i = \text{the effect of the } i^{th} \text{ Brand,}$$
$$\beta_{(i)j} = \text{the effect of the } j^{th} \text{ Administration Method of the } i^{th} \text{ Brand,}$$
$$E_{ijk} = \text{the random error.}$$

(c) The following ANOVA table was found using SAS and PROC ANOVA.

Analysis of Variance Procedure

Dependent Variable: BSUGAR

Source	DF	Sum of Squares	Mean Square	F Value	Pr > F
Model	5	683.61111	136.72222	22.79	0.0001
Error	12	72.00000	6.00000		
Corrected Total	17	755.61111			

R-Square	C.V.	Root MSE	BSUGAR Mean
0.904713	10.99522	2.4495	22.278

Source	DF	Anova SS	Mean Square	F Value	Pr > F
BRAND	2	611.44444	305.72222	50.95	0.0001
ADMIN(BRAND)	3	72.16667	24.05556	4.01	0.0344

The first thing to do is test the nested factor. Since the nested factor has a p-value of .0344, less than most α-levels, there is some evidence that method of administration within Brand is having a significant effect. The next step in the analysis would be to test the individual administration methods within each brand itself.

(d) Because the nested factor is significant ($p = 0.0344$), we should look at the simple effects of administration for each brand. The next step in the analysis would be to test the individual administration methods within each brand itself.

(e) Using the notation in the text we get

$$\hat{\mu}[A_1 B] = \bar{y}_{(1)2} - \bar{y}_{(1)1} = \frac{59}{3} - \frac{47}{3} = 4.0$$
$$\hat{\mu}[A_2 B] = \bar{y}_{(2)2} - \bar{y}_{(2)1} = \frac{52}{3} - \frac{60}{3} = -2.6667$$
$$\hat{\mu}[A_3 B] = \bar{y}_{(3)2} - \bar{y}_{(3)1} = \frac{99}{3} - \frac{84}{3} = 5.0.$$

Each mean is calculated based on 3 observations and the MS[E]=6.0 with 12 df. The 90% simultaneous confidence intervals are constructed using the Bonferroni approach.

The intervals have the following form

$$\hat{\mu}[A_iB] \pm t(12, \tfrac{0.10}{6})\sqrt{MS[E]\left(\frac{1}{3} + \frac{1}{3}\right)}.$$

For the simple effect of method of administration for Brand A this becomes

$$4.0 \pm 2.4030(2.0) \quad \text{or} \quad (-0.806, 8.806).$$

For the simple effect of method of administration for Brand B this becomes

$$-2.6667 \pm 2.4030(2.0) \quad \text{or} \quad (-7.4727, 2.1393).$$

For the simple effect of method of administration for Insulin this becomes

$$5.0 \pm 2.4030(2.0) \quad \text{or} \quad (0.194, 9.806).$$

Thus we conclude that with 90% confidence that simultaneously the simple effects of method of administration all fall within their respective intervals. Based on this it seems that only method of administration differs for Insulin.

(f) The comparisons are to be made on the main effects of level A, Drug type. The Tukey comparison method compares each absolute mean difference to the value

$$q(t, dferror, \alpha)\sqrt{\frac{MS[E]}{n}} = q(3, 12, 0.05)\sqrt{\frac{6}{6}} = 3.77(1) = 3.77.$$

The individual comparisons can be made using the following table.

Comparison	$\bar{Y}_3 - \bar{Y}_2$	$\bar{Y}_3 - \bar{Y}_1$	$\bar{Y}_2 - \bar{Y}_1$
Diff	30.5-18.6667	30.5-17.6667	18.6667-17.6667
Abs. Diff.	11.8333	12.8333	1.0
Significance	S	S	NS

Based on this result and those made above, there seems to be a large difference between the effects of Insulin and the other drugs. Also within Insulin the method of administration seems to be quite important. There does not seem to be any evidence of differences for either drugs A or B or within them an effect of method of administration.

Chapter 14

Random and Mixed Effects Anova

14.1. (a) A one-way ANOVA model for the data would be

$$Y_{ij} = \mu + T_i + E_{ij} \qquad i = 1, 2, \ldots, 10; j = 1, 2, 3,$$

where

Y_{ij} = percent protein content of the seeds produced by the ith F_2 generation plant in the jth plot.

μ = overall mean percent protein of the seeds produced by all plants,

T_i = effect of the ith randomly selected plant, the T_i are independent $N(0, \sigma_T^2)$, and

E_{ij} = random error. E_{ij} are independent $N(0, \sigma^2)$, the T_i and E_{ij} are independent.

(b) A constant protein content is equivalent to no variability in the protein content. That is H_0: $\sigma_T^2 = 0$.

(c) H_0: $\mu = 40$.

(d) σ_E^2 measures the variability among the experimental units, that is the variability in the measured protein contents of F_3 plants grown from the seeds of a single F_2 plant.

14.3. (a)

$$Y_{ij} = \mu + T_i + E_{ij} \qquad i = 1, 2, 3, 4, 5; j = 1, 2, 3, 4,$$

where

Y_{ij} = cation-exchange capacity of the jth sample from the ith core,

μ = overall mean cation-exchange capacity of all core samples,

T_i = effect of the ith randomly selected core. The T_i are independent $N(0, \sigma_T^2)$,

E_{ij} = random error. E_{ij} are independent $N(0, \sigma^2)$ and T_i and E_{ij} are independent.

(b) H_0: $\mu > 18$

14.5. (a) By running the data through SAS using PROC Anova, we can put the following partial ANOVA table together. The expected mean squares are calculated as in the text. Since all sample sizes are equal $n_0 = n = 3$.

Source	DF	Mean Square	F Value	Pr > F	EMS
Plant	9	8.223556	4.25	0.0034	$\sigma^2 + 3\sigma_T^2$
Error	20	1.935333			σ^2
Corrected Total	29				

(b) The associated p-value of this test is .0034. Since this p-value is smaller than almost all α-levels used in practice, there is sufficient evidence to conclude that there is a significant amount of plant to plant variation.

(c) Following the arguments in the text, we get

$$\hat{\sigma}^2 = 1.9353,$$

and

$$\hat{\sigma}_T^2 = \frac{8.2236 - 1.9353}{3} = 2.0961.$$

From this we see that plant to plant variation is about the same as within plant variation.

(d) Firstly $\overline{Y}_{++} = 39.4267$, so that

$$\widehat{CV} = \frac{\sqrt{\hat{\sigma}^2 + \hat{\sigma}_T^2}}{\overline{Y}_{++}} = \frac{\sqrt{1.9353 + 2.0961}}{39.4267} = \frac{2.0078}{39.4267} = .0509.$$

It follows that the standard deviation of protein content is close to 5% of the mean protein content.

(e) Following the notation in the text, a 95% confidence interval for ρ_1 requires the following:

$$F_c = 4.25, \quad F_{.975} = F(9, 20, .975) = .2727 \quad \text{and} \quad F_{.025} = 2.84.$$

Thus, since $n = 3$, a 95% confidence interval for ρ_1 is

$$\left(\frac{4.25 - 2.84}{4.25 + (2)2.84}, \frac{4.25 - .2727}{4.25 + (2).2727} \right) \quad \text{or} \quad (.1420, .8294).$$

With 95% confidence we conclude that between 14.20% and 82.94% of the variation in measured protein content is due to differences between F_2 generation plants.

(f) A test on the average protein content of seeds produced in the F_3 generation is a test about μ, the overall mean. An estimate of μ is $\hat{\mu} = \overline{Y}_{++} = 39.4267$. The associated standard error is

$$\hat{\sigma}_{\hat{\mu}} = \sqrt{\frac{MS[T]}{30}} = \sqrt{\frac{8.2236}{30}} = .5236.$$

The test can be formed as follows:

H$_0$: $\mu \leq 40$

H$_1$: $\mu > 40$

Test Statistic: $t_c = \dfrac{39.4267 - 40}{.5236} = -1.0949,$
which is compared to a t_9.

Conclusion: The associated p-value is .8490, so the null hypothesis is not rejected. There is insufficient evidence that the average protein content of seeds produced in the F_3 generation is more than 40%. This should be obvious since the point estimate is less than 40.

14.7. (a) By running the data through SAS using PROC ANOVA, we get the following ANOVA table. (Note: PROC ANOVA in SAS is okay here because of the one-way classification. With a more complicated model, the unequal subclass numbers would require the use of PROC GLM.)

```
Dependent Variable: CAT_EXC
                                 Sum of          Mean
Source                    DF     Squares         Square    F Value    Pr > F
Model                      4     9.4413235     2.3603309      2.06    0.1488
Error                     12    13.7175000     1.1431250
Corrected Total           16    23.1588235

             R-Square           C.V.       Root MSE         CAT_EXC Mean
             0.407677        5.558377        1.0692               19.235

Source                    DF    Anova SS   Mean Square    F Value    Pr > F
CORE                       4    9.4413235    2.3603309       2.06    0.1488
```

The test that Cation–exchange capacity does not vary over the region has associated p–value $= 0.1488$. Since this is larger than all α–levels used in practice, there is little evidence to conclude that Cation–exchange capacity varies over the region.

(b) Since there is some missing data we must calculate n_0.

$$n_0 = \frac{1}{4}\left(17 - \frac{5^2 + 5^2 + 4^2 + 3^2}{17}\right) = 3.1471$$

Based on this we get the following two estimates of the variance components:

$$\hat{\sigma}^2 = MS[E] = 1.1431$$
$$\hat{\sigma}_T^2 = \frac{MS[T] - MS[E]}{n_0} = \frac{2.3603 - 1.1431}{3.1471} = 0.3868$$

The estimated variance component due to differences between cores is about a third of the estimated variance component due to differences between parts in the same core.

(c) The estimate of mean Cation–exchange capacity for the region is just $\overline{Y}_{++}$.

$$\hat{\mu} = \overline{Y}_{++} = \frac{327}{17} = 19.2353.$$

(d) Note that

$$\overline{Y}_{++} = \frac{\sum_i \sum_j (\mu + T_i + E_{ij})}{17} = \mu + \frac{4T_1 + 4T_2 + 3T_3 + 2T_4 + 4T_5}{17} + \frac{\sum_i \sum_j E_{ij}}{17}.$$

By the assumptions of independence of T_i and E_{ij} and the usual variance properties we find that

$$\sigma_{\overline{Y}_{++}}^2 = \left(\frac{4^2}{17^2} + \frac{4^2}{17^2} + \frac{3^2}{17^2} + \frac{2^2}{17^2} + \frac{4^2}{17^2}\right)\sigma_T^2 + \frac{\sigma^2}{17}$$
$$= \frac{61\sigma_T^2}{17^2} + \frac{\sigma^2}{17} = \frac{61\sigma_T^2 + 17\sigma^2}{17^2}.$$

In general we have the result

$$\sigma_{\overline{Y}_{++}}^2 = \frac{\sum_i n_i^2 \sigma_T^2 + N\sigma^2}{N^2}.$$

The appropriate estimated standard error is

$$\hat{\sigma}_{\overline{Y}_{++}} = \sqrt{\frac{61\hat{\sigma}_T^2 + 17\hat{\sigma}^2}{17^2}} = \sqrt{\frac{61(0.3868) + 17(1.1431)}{17^2}} = 0.3859.$$

(e) The only difficulty that one would have in constructing an upper bound for this Exericse is the appropriate degrees of freedom. i.e. the general formulation for the upper bound would be

$$\hat{\mu} + t(df, 0.05)\hat{\sigma}_{\hat{\mu}}.$$

A point estimate of μ is $\overline{Y}_{++}$ and the standard error was calculated above. The difficulty is that the standard error is based on a weighted function of Mean Squares and, therefore, an extension of Satterthwaite's procedure (Section 4.3) would need to be applied. Using this extension, which is discussed in Box 14.9 as the Satterthwaite-Welch method, the lower bound is 20.1435 meq/100g.

14.9. (a) Based on the information given we want to use the hypothesis testing approach with $\alpha = .10$, $\beta = .20$ and $\delta = 2$. Using statistical software we arrive at the following table.

t	n	ndf	ddf	F	fdelt	Power
2	2	1	2	18.512821	3.7025641	0.1942205
2	3	1	4	7.708647	1.1012353	0.3532135
2	4	1	6	5.987378	0.6652642	0.4458706
2	5	1	8	5.317655	0.4834232	0.5065657
3	2	2	3	9.552094	1.9104189	0.2916925
3	3	2	6	5.143253	0.7347504	0.5182974
3	4	2	9	4.256495	0.4729439	0.6378159
3	5	2	12	3.885294	0.3532085	0.7094942
4	2	3	4	6.591382	1.3182764	0.3850512
4	3	3	8	4.066181	0.5808829	0.6439196
4	4	3	12	3.490295	0.3878105	0.7638635
4	5	3	16	3.238872	0.2944429	0.8288344
5	2	4	5	5.192168	1.0384336	0.4706658
5	3	4	10	3.478050	0.4968642	0.7388770
5	4	4	15	3.055568	0.3395076	0.8470813

The table shows that using $(t, n) = (4, 5)$, or $(5, 4)$ gives the desired power of at least 0.80. The decision of which to use depends on which costs more treatments or replications.

(b) Recalling that from Exercise **14.5** we have the following estimates

$$\hat{\sigma}^2 = 1.9353 \qquad \text{and} \qquad \hat{\sigma}_T^2 = 2.0961.$$

The desired half length of the interval (or margin of error) is 5. Hence we need to find t and n to satisfy

$$t \geq F(1, t-1, 0.10)\left[\left(\frac{1.9353}{5^2}\right) + \frac{1}{n}\left(\frac{2.0961}{5^2}\right)\right]$$

$$\geq F(1, t-1, 0.10)\left[0.0774 + \frac{0.0838}{n}\right].$$

The following table shows that taking 3 plants with two plots per plant will satisfy our constraints.

t	2	n 3	4	5
2	4.7565	4.1995	3.9210	3.7539
3	1.0174			

(c) We have estimates of σ^2 and σ_T^2 from above. We also note that $C_0 = 150$ and $C_1 = 50$. We also require that the total cost, C, not exceed 2,000. By plugging these into the formulas in the text we find that

$$n = \sqrt{\frac{C_0 \hat\sigma^2}{C_1 \hat\sigma_T^2}} = \sqrt{\frac{50(2.0961)}{150(1.9353)}} = 1.8026 \sim 2.$$

Plugging this in to solve for t gives

$$t = \frac{2000}{150 + 50(2)} = 8.$$

Taking 8 plants each with only two plots per plant satisfies the constraint. In fact, the total cost of conducting such a study is exactly 2,000.

(d) The standard error of $\overline{Y}_{++}$ can be estimated as

$$\hat\sigma_{\overline{Y}_{++}} = \sqrt{\frac{MS[T]}{nl}} = \sqrt{\frac{8.2236}{2(8)}} = 0.7169.$$

The width of a 95% confidence interval for μ will be approximately equal to $2t(7, 0.025)(0.7169) = 3.4$. Thus, μ can be estimated with error less than 1.7.

14.11. The restriction is that the estimated variance of the overall mean is less than 0.2. The variance of the overall mean is

$$\sigma_{\overline{Y}_{++}} = \frac{\sigma^2}{nt} + \frac{\sigma_T^2}{t}.$$

From Exercise **14.7** we have that

$$\hat\sigma^2 = 1.1431 \qquad \text{and} \qquad \hat\sigma_T^2 = 0.3868.$$

Therefore, we are looking for values of (n, t) such that

$$0.2 \le \frac{1.1431}{nt} + \frac{0.3868}{t}.$$

This can be programmed in SAS to get the following table.

			n		
t	2	3	4	5	6
3	0.3194	0.2559	0.2242	0.2051	0.1924
4	0.2396	1.1919			
5	0.1917				

Hence, values for (n, t) of $(6, 3)$, $(4, 4)$, and $(5, 2)$ are all acceptable. $(5, 2)$ would keep the total number of observations as small as possible.

14.13. The 2–way mixed model with crossed factors is

$$Y_{ijk} = \mu + \alpha_i + B_j + (\alpha B)_{ij} + E_{ijk}.$$

The expected difference between Y_{1jk} and Y_{2jk} is

$$
\begin{aligned}
\mathrm{E}(Y_{1jk} - Y_{2jk}) &= \mathrm{E}(\mu + \alpha_1 + B_j + (\alpha B)_{1j} + E_{1jk} - \mu - \alpha_2 - B_j - (\alpha B)_{2j} - E_{2jk}) \\
&= \mathrm{E}(\alpha_1 - \alpha_2 + (\alpha B)_{1j} - (\alpha B)_{2j} + E_{1jk} - E_{2jk}) \\
&= \alpha_1 - \alpha_2 + \mathrm{E}(\alpha B)_{1j} - \mathrm{E}(\alpha B)_{2j} + \mathrm{E}(E_{1jk}) - \mathrm{E}(E_{2jk}) \\
&= \alpha_1 - \alpha_2.
\end{aligned}
$$

Note: in the last step we have used the fact that $\mathrm{E}(\alpha\beta)_{ij}$, etc. are zero.j

161

14.15. (a) By using the hint and the example in the text we see that

$$a = 3, \quad b = 4, \quad n = 3, \quad c_1 = \frac{1}{2}, \quad c_2 = \frac{1}{2}, \quad c_3 = -1.$$

A point estimate of θ is

$$\hat{\theta} = \frac{1}{2}(\overline{Y}_{1++} + \overline{Y}_{2++}) - \overline{Y}_{3++} = \frac{1}{2}(16.76 + 17.92) - 15.92 = 1.42.$$

To find the standard error of $\hat{\theta}$ we need $\sum c_i = 0$ and $\sum c_i^2 = 1.5$. From the ANOVA table in the text we get $MS[B] = 4.6046$ and $MS[AB] = 0.1554$. Next,

$$\sigma_{\hat{\theta}}^2 = \frac{3\sigma_{\alpha\beta}^2 + \sigma^2}{3(4)}\left(\sum c_i^2\right)$$

$$\hat{\sigma}_{\hat{\theta}}^2 = \frac{MS\,[AB]}{3(4)}(1.5)$$

$$= \frac{4.6046(1.5)}{12} = 0.5756$$

The degrees of freedom for the interval are those associated with MS[AB], i.e. 6. The 95% confidence interval for θ is

$$1.42 \pm t(6, 0.025)(0.7587) = 1.42 \pm 2.4469(0.7587) \quad \text{or} \quad (0.45, 3.29).$$

With 95% confidence we conclude that the mean response for Fertilization methods 1 and 2 is between 0.45 and 3.29 higher than the mean response for Fertilization method 3.

(b) For all of the individual differences between $\overline{Y}_{i++}$ and $\overline{Y}_{j++}$ the estimated standard errors and degrees of freedom for error are the same as in the example in the text, i.e.

$$\nu = 16 \quad \text{and} \quad \hat{\sigma}_{\overline{Y}_{i++} - \overline{Y}_{j++}} = 0.1228.$$

There are a total of three distinct treatments. By using the Bonferroni approach each interval takes the general form

$$\overline{Y}_{i++} - \overline{Y}_{j++} \pm t(16, \tfrac{0.05}{6})(\hat{\sigma}_{\overline{Y}_{i++} - \overline{Y}_{j++}}).$$

For Fertilizer methods 1 and 2 this becomes

$$16.76 - 17.92 \pm 2.6730(0.1228) \quad \text{or} \quad (-1.1882, -0.8318).$$

For Fertilizer methods 1 and 3 this becomes

$$16.76 - 15.92 \pm 2.6730(0.1228) \quad \text{or} \quad (0.5118, 1.1682).$$

For Fertilizer methods 2 and 3 this becomes

$$17.92 - 15.92 \pm 2.6730(0.1228) \quad \text{or} \quad (1.6718, 2.3282).$$

With at least 95% confidence we conclude that simultaneously the expected mean differences between each pair of fertilizers is within their respective interval.

14.17. (a) Let Y_{ijk} denote the radon flux measurement for the ith Method, jth Technician, and kth plot. Then

$$Y_{ijk} = \mu + \alpha_i + B_{(i)j} + E_{ijk} \quad i = 1, 2; j = 1, \ldots, 6; k = 1, 2,$$

where

$$\mu = \text{overall mean measured radon flux value for all plots}$$

$$\alpha_i = \text{the fixed effect of the } i^{th} \text{ method, } \alpha_1 + \alpha_2 = 0,$$

$$B_{(i)j} = \text{the random effect of the } j^{th} \text{ technician within the } i^{th} \text{ method. The } B_{(i)j}\text{'s}$$
are assumed ind. $N(0, \sigma^2_{B(A)})$,

$$E_{ijk} = \text{the random error. The } E_{ijk}\text{'s are assumed ind. } N(0, \sigma^2) \text{ and the } B_{(i)j}\text{'s}$$
and E_{ijk}'s are mutually independent.

(b) The following ANOVA table can be constructed using the computing formulas in the text or by using SAS.

Source	df	SS	MS	F_c	$P > F$
Method(A)	1	7.5937	7.5937	1.59	0.2362
Tech(Method) B(A)	10	47.8108	4.7811	7.49	0.0009
Error	12	7.6550	0.6379		
Total	23	63.0596			

Based on the table there is a significant amount of variability between Technicians within the same Method (p−value = 0.0009). There does not seem to be a difference between the two methods (p−value = 0.2362).

(c) Following the formulas in the text we find that

$$\hat{\sigma}^2 = MS[E] = 0.6379,$$

$$\hat{\sigma}^2_{B(A)} = \frac{MS[B(A)] - MS[E]}{n} = \frac{4.7811 - 0.6379}{2} = 2.0716.$$

Since $\hat{\sigma}^2_{B(A)}$ is approximately three times as large as $\hat{\sigma}^2$, there seems to be approximately three times as much variability between measurement of different technicians within a specific method as between measurements by the same technician.

(d) $\overline{Y}_{1++}$ is an estimate of the expected radon flux measurement resulting from Method 1. This can most easily be seen by taking the expectation of $\overline{Y}_{1++}$. Since both $B_{(i)j}$ and E_{ijk} are assumed to be random quantities with zero means, they have expectation zero. Hence all that is left is $\mu + \alpha_1$, the expected radon flux measurement resulting from Method 1.

(e) The statement in the text is equivalent to showing that the variance of $\hat{\theta}$, denoted $\text{Var}(\hat{\theta})$, is equal to $\frac{E(MS[B(A)])}{nb}$. To do this, first notice that $\overline{Y}_{1++}$ equals a constant plus the sum of the means of two independent samples, one of size b and another of size nb. Then, we have

$$\begin{aligned}
\text{Var}(\overline{Y}_{1++}) &= \text{Var}\left(\mu + \alpha_1 + \frac{\sum_j B_{(1)j}}{b} + \frac{\sum_j \sum_k E_{1jk}}{nb}\right) \\
&= \sum_j \text{Var}\left(\frac{B_{(1)j}}{b}\right) + \sum_j \sum_k \text{Var}\left(\frac{E_{1jk}}{nb}\right) \\
&= \sum_j \frac{\sigma^2_{B(A)}}{n^2} + \sum_j \sum_k \frac{\sigma^2}{(nb)^2} \\
&= \frac{n\sigma^2_{B(A)} + \sigma^2}{nb} = \frac{E(MS[B(A)])}{nb}.
\end{aligned}$$

163

Hence a natural estimate is $\frac{MS[B(A)]}{nb}$. Following the notation in Box 14.8 we have $c_1 = 1$, hence $\sum c_i^2 = 1$, and

(f) The are 10 degrees of freedom associated with $MS[B(A)]$ so that a 99% confidence interval for θ is

$$\overline{Y}_{1++} \pm t(10, 0.005) \sqrt{\frac{MS[B(A)]}{nb}}.$$

We have that $n = 2$, $b = 6$, $MS[B(A)] = 4.7811$, and $\overline{Y}_{1++}$ can be calculated as

$$\overline{Y}_{1++} = \frac{42.4 + \ldots 42.3}{12} = 41.4083.$$

Thus, the confidence interval takes the form

$$41.4083 \pm 3.1693(0.6312) \quad \text{or} \quad (39.0825, 43.083).$$

With 99% confidence we conclude that the expected radon flux measurement for Method 1 is between 39.0825 and 43.083.

(g) Using the results in Box 14.8, the standard error of the difference between the two means can be shown to be

$$\hat{\sigma}_{\overline{Y}_{1++} - \overline{Y}_{2++}} = \sqrt{\frac{2MS[B(A)]}{nb}}.$$

The upper bound is

$$\overline{Y}_{1++} - \overline{Y}_{2++} + t(10, 0.05) \sqrt{\frac{MS[B(A)]}{6}}.$$

By calculating $\overline{Y}_{2++} = 40.2833$ and plugging into the above formula we get the bound as

$$41.4083 - 40.2833 + 1.8125(0.8927) = 2.7430.$$

With 95% we conclude that the difference between the expected radon flux measurements for Methods 1 and 2 is at most 2.7430.

14.19. (a) Let Y_{ijk} denote the amylase value for the ith lab, j^{th} technician, and k^{th} specimen. Then

$$Y_{ijk} = \mu + \alpha_i + B_{(i)j} + E_{ijk} \quad i = 1, \ldots, 4; j = 1, \ldots, 4; k = 1, \ldots, 10,$$

where

μ	=	overall mean measured amylase value for all specimens,
α_i	=	the fixed effect of the ith lab, $\alpha_1 + \ldots + \alpha_4 = 0$,
$B_{(i)j}$	=	the random effect of the jth technician within the ith lab. The $B_{(i)j}$'s are assumed ind. $N(0, \sigma^2_{B(A)})$,
E_{ijk}	=	the random error. The E_{ijk}'s are assumed ind. $N(0, \sigma^2)$ and the $B_{(i)j}$'s and E_{ijk}'s are mutually independent.

(b) We can calculate the sample means for each lab.

$$\overline{Y}_{1++} = \frac{152}{40} = 3.8 \quad \overline{Y}_{2++} = \frac{144}{40} = 3.6 \quad \overline{Y}_{3++} = \frac{184}{40} = 4.6 \quad \overline{Y}_{4++} = \frac{176}{40} = 4.4$$

Any of the pairwise comparison methods in Section 9.5 can be used. Since the variance of $\overline{Y}_{i++}$ is estimated by $MS[B(A)]/nb$, we replace n, MS[E], and ν in Section 9.5 with

$nb = 12$, MS[B(A)] $= 1.63$, and df for MS[B(A)] $= 12$, respectively. With this, the standard error of a difference is

$$\hat{\sigma}_{\overline{Y}_{i++} - \overline{Y}_{j++}} = \sqrt{\frac{(2)MS[B(A)]}{nb}} = \sqrt{\frac{2(1.63)}{10(4)}} = 0.2855.$$

The multiple comparison procedure we will use is the Bonferroni approach, with $\alpha = .05$. Each absolute difference between means is compared to the critical contrast value of

$$t(12, \tfrac{0.05}{12})(0.2855) = 3.1527(0.2855) = 0.9001.$$

The following table makes the appropriate comparisons.

Labs	Diff	Abs. Diff	Conclusion
1 vs. 2	$3.8 - 3.6$	0.20	NS
1 vs. 3	$3.8 - 4.6$	0.80	NS
1 vs. 4	$3.8 - 4.4$	0.60	NS
2 vs. 3	$3.6 - 4.6$	1.00	S
2 vs. 4	$3.6 - 4.4$	0.80	NS
3 vs. 4	$4.6 - 4.4$	0.20	NS

With 95% confidence we conclude that only differences between labs 2 and 3 exist. There is no evidence of differences between other labs.

(c) We need to find estimates of σ^2 and $\sigma^2_{B(A)}$. Following the formula in the text we have

$$\hat{\sigma}^2 = MS[E] = 1.02,$$
$$\hat{\sigma}^2_{B(A)} = \frac{MS[B(A)] - MS[E]}{n} = \frac{1.63 - 1.02}{10} = 0.061.$$

This gives the obvious estimate

$$\hat{\rho} = \frac{\hat{\sigma}^2_{B(A)}}{\hat{\sigma}^2_{B(A)} + \hat{\sigma}^2} = \frac{0.061}{0.061 + 1.02} = 0.0564.$$

Since $\hat{\rho}$ is small it implies that the amount of variability in measurement by the same technician is small in comparison to the amount of variability in measurements by the different technicians. Approximately 5-6% of the variability in the measurement of serum amylase is due to variability within technicians.

(d) By substituting in the sample means for the population means we get the following point estimates.

$$\hat{\theta}_1 = \overline{Y}_{1++} - \overline{Y}_{2++} = 3.8 - 3.6 = .2$$
$$\hat{\theta}_2 = \frac{1}{2}(\overline{Y}_{1++} + \overline{Y}_{2++}) - \frac{1}{2}(\overline{Y}_{3++} + \overline{Y}_{4++}) = \frac{1}{2}(3.55 + 3.6) - \frac{1}{2}(4.6 + 4.4) = -.925$$
$$\hat{\theta}_3 = \overline{Y}_{3++} - \overline{Y}_{4++} = 4.6 - 4.4 = .2$$

The appropriate standard errors for can be found using Box 14.8. These calculations give

$$\hat{\sigma}_{\hat{\theta}_1} = \hat{\sigma}_{\hat{\theta}_3} = 0.2855, \quad \text{and} \quad \hat{\sigma}_{\hat{\theta}_2} = 0.2019.$$

The standard errors of all three contrasts depend only on $MS[B(A)]$ so the appropriate degrees of freedom remains 12. Using the Bonferroni approach each interval takes the form

$$\hat{\theta}_i \pm t(12, \tfrac{0.10}{6})\hat{\sigma}_{\hat{\theta}_i}.$$

165

This gives for θ_1

$$0.2 \pm 2.4030(.2855) \quad \text{or} \quad (-0.4861, 0.8861).$$

This gives for θ_2

$$-0.925 \pm 2.4030(0.2019) \quad \text{or} \quad (-1.4102, -0.4398).$$

This gives for θ_3

$$0.2 \pm 2.4030(0.2855) \quad \text{or} \quad (-0.4861, 0.8861).$$

With 90% confidence we conclude that simultaneously the expected values of the contrasts lie within their respective intervals.

14.21. (a) $\text{Var}(\overline{Y}_{+++}) = \text{Var}(\mu + \overline{T}_+ + \overline{R}_{++} + \overline{S}_{+++}) = \text{Var}(\overline{T}_+) + \text{Var}(\overline{R}_{++}) + \text{Var}(\overline{S}_{+++})$ where

$$\text{Var}(\overline{T}_+) = \frac{\sigma_T^2}{t}$$

$$\text{Var}(\overline{R}_{++}) = \frac{\sigma_R^2}{ts_1}$$

$$\text{Var}(\overline{S}_{+++}) = \frac{\sigma_S^2}{ts_1 s_2}$$

Then

$$\text{Var}(\overline{Y}_{+++}) = \frac{\sigma_T^2}{t} + \frac{\sigma_R^2}{ts_1} + \frac{\sigma_S^2}{ts_1 s_2} = \frac{\sigma_S^2 + s_1 \sigma_R^2 + s_1 s_2 \sigma_T^2}{ts_1 s_2}.$$

(b) Since the expected mean square for A is

$$EMS[A] = \sigma_S^2 + s_1 \sigma_R^2 + s_1 s_2 \sigma_T^2$$

an estimator for the standard error of $\overline{Y}_{+++}$ is

$$\hat{\sigma}_{\overline{Y}_{+++}} = \sqrt{\frac{MS[A]}{ts_1 s_2}}.$$

since $MS[A]$ estimates the appropriate linear combination of variances as shown in the text.

14.23. For a 2-way ANOVA model with fixed effects we use the model

$$Y_{ijk} = \mu + \alpha_i + \beta_j + \alpha\beta_{ij} + E_{ijk} \qquad i = 1, 2, \ldots, a; j = 1, 2, \ldots, b; k = 1, 2, \ldots, n.$$

First, consider $EMS[A]$. We start by treating A as random. It is a linear combination of each variance component that corresponds to an effect whose subscripts are subscripts of A. The only other variance component is the error variance. Therefore, $EMS[A]$ is a linear combination of σ_A^2 and σ^2. The coefficients are bn, the ranges for j and k, and 1, for the error variance. This gives us

$$EMS[A] = bn\sigma_A^2 + \sigma^2.$$

Finally, we replace the variance component of A with its effect size. This gives

$$EMS[A] = bn\psi_A^2 + \sigma^2.$$

Using the same procedure the other expected mean squares are

$$EMS[B] = an\psi_B^2 + \sigma^2,$$
$$EMS[AB] = n\psi_{AB}^2 + \sigma^2, \quad \text{and}$$
$$EMS[E] = \sigma^2.$$

For a 2-way mixed model with crossed effects, A fixed and B random, we have

$$EMS[A] = bn\psi_A^2 + n\sigma_{\alpha B}^2 + \sigma^2,$$
$$EMS[B] = an\sigma_B^2 + n\sigma_{\alpha B}^2 + \sigma^2,$$
$$EMS[AB] = n\sigma_{\alpha B}^2 + \sigma^2, \quad \text{and}$$
$$EMS[E] = \sigma^2.$$

The others are done similarly.

Chapter 15

ANOVA Models with Block Effects

15.1. (a) Assuming random block effects and fixed treatment effects, the additive model for an RCB design is

$$Y_{ij} = \mu + R_i + \tau_j + E_{ij} \qquad i = 1, 2, \ldots, t; j = 1, 2, \ldots, b$$

Then,

$$\overline{Y}_{+j} = \frac{1}{t} \sum_{i=1}^{t} Y_{ij} = \mu + \frac{1}{t} \sum_{i=1}^{t} R_i + \tau_j + \frac{1}{t} \sum_{i=1}^{t} E_{ij}$$

$$= \mu + \overline{R}_+ + \tau_j + \overline{E}_{+j}.$$

A contrast of treatment means is

$$\hat{\theta} = c_1 \overline{Y}_{+1} + c_2 \overline{Y}_{+2} + c_3 \overline{Y}_{+3} + \cdots + c_t \overline{Y}_{+t}$$

$$= c_1(\mu + \overline{R}_+ + \tau_1 + \overline{E}_{+1}) + c_2(\mu + \overline{R}_+ + \tau_2 + \overline{E}_{+2}) + \cdots + c_t(\mu + \overline{R}_+ + \tau_t + \overline{E}_{+t})$$

$$= (\mu + \overline{R}_+)(c_1 + c_2 + \cdots + c_t) + (c_1\tau_1 + c_2\tau_2 + \cdots + c_t\tau_t) + (c_1\overline{E}_{+1} + c_2\overline{E}_{+2} + \cdots + c_t\overline{E}_{+t})$$

$$= (c_1\tau_1 + c_2\tau_2 + \cdots + c_t\tau_t) + (c_1\overline{E}_{+1} + c_2\overline{E}_{+2} + \cdots + c_t\overline{E}_{+t})$$

since $\sum c_i = 0$. Note that this result is true for either fixed or random blocks.
Similar steps show the result for random treatment effect.

(b) Consider the result from Part a. $c_1\tau_1 + c_2\tau_2 + c_3\tau_3 + \cdots + c_t\tau_t$ is a constant that is the value of the contrast among population means and $c_1\overline{E}_{+1} + c_2\overline{E}_{+2} + c_3\overline{E}_{+3} + \cdots + c_t\overline{E}_{+t}$ is the linear combination of the means of independent samples from normal populations.

15.3. (a)

$$Y_{ij} = \mu + R_i + \tau_j + E_{ij} \qquad i = 1, 2, 3, \ldots, 6; j = 1, 2, 3, 4.$$

where

Y_{ij} = the serum amylase value of the ith patient determined by the jth method,
μ = overall mean serum amylase,
R_i = random effect of the ith patient, the R_i are independent $N(0, \sigma_R^2)$,
τ_j = the fixed effect of the jth method, $\sum \tau_j = 0$, and
E_{ij} = random error in the serum amylase of the ith patient and the jth method. The E_{ij} are independent $N(0, \sigma^2)$ and are independent of the R_i.

(b) The ANOVA table is (from SAS)

Analysis of Variance Procedure

Dependent Variable: VALUE

Source	DF	Sum of Squares	Mean Square	F Value	Pr > F
PATIENT	5	2157788.3	431557.7	263.21	0.0001
METHOD	3	116979.3	38993.1	23.78	0.0001
Error	15	24593.7	1639.6		
Corrected Total	23	2299361.3			

The F-tests are

H_0: $\psi_T^2 = 0$

H_1: $\psi_T^2 \neq 0$

Test Statistic: $F_c = \dfrac{MS[T]}{MS[E]} = 23.78$

Conclusion: We reject H_0. There is sufficient evidence (p−value = 0.0001) to say that a significant Method effect exists.

H_0: $\sigma_R^2 = 0$

H_1: $\sigma_R^2 \neq 0$

Test Statistic: $F_c = \dfrac{MS[R]}{MS[E]} = 263.21$

Conclusion: We reject H_0. There is sufficient evidence (p−value = 0.0001) to say that a significant Patient effect exists.

(c) The estimate of the difference between the true mean serum amylase values determined using Methods 1 and 4 is

$$\hat{\theta} = \overline{Y}_{+1} - \overline{Y}_{+4} = 700.3333 - 854.6667 = -154.3334$$

The t-statistic is $t((n-1)(r-1), \alpha/2) = t(15, 0.025) = 2.131$ and $c_1 = 1$, $c_2 = 0$, $c_3 = 0$, $c_4 = -1$. Then, the standard error is

$$\hat{\sigma}_{\hat{\theta}} = \sqrt{\frac{2}{r}MS[E]} = \sqrt{\frac{2}{6}(1639.6)} = 23.3780.$$

The 95% confidence interval is

$$\hat{\theta} \pm t((n-1)(r-1), \alpha/2)\hat{\sigma}_{\hat{\theta}} = -154.33 \pm (2.131)(23.3780)$$

or $(-204.1, -104.5)$. With 95% confidence we conclude that the difference between the true mean serum amylase values determined using Methods 1 and 4 is between −204.1 and −104.5. Or that the expected serum amylase value determined using Method 1 is between 204.1 and 104.5 less than the expected serum amylase value determined using Method 4.

(d) No, it would not be reasonable to conclude that the two methods yield the same values on the average, because the above interval does not include zero.

(e) 90% simultaneous upper bounds for the population mean values for Methods 1 and 3 can be calculated similar to Example 15.4. The standard error of $\overline{Y}_{+j}$ is estimated as

$$\hat{\sigma}_{\overline{Y}_{+j}} = \sqrt{\frac{1}{rt}MS[R] + \frac{1}{r}MS[E]} = \sqrt{\frac{431557.7}{24} + \frac{1639.6}{6}} = 135.1105.$$

Because the variance is a linear combination of mean squares, the degrees of freedom is calculated with the Satterthwaite-Welch method of Box 14.9.

$$\nu^* = \frac{\left(\left(\frac{1}{24}\right)431557.7 + \left(\frac{1}{8}\right)1639.6\right)^2}{\left(\frac{1}{24}\right)^2 \frac{431557.7^2}{5} + \left(\frac{1}{8}\right)^2 \frac{1639.6^2}{15}} = 5.1144.$$

Hence, $\nu = 5$. Then, the bounds are (with $\alpha = .10$)

$$\overline{Y}_{+1} + t(5, 0.025)\hat{\sigma}_{\overline{Y}_{+1}} = 700.3333 + (2.5706)(135.1105) = 1047.6483$$

and

$$\overline{Y}_{+3} + t((5, 0.025)\hat{\sigma}_{\overline{Y}_{+1}} = 697 + (2.5706)(135.1105) = 1044.3150.$$

Since both upper bounds are greater than 750, we can conclude that Methods 1 and 3 both produce on the average values greater than 750 as well as less than 750.

(f) The serum amylase values will be more similar for samples from a single patient than for samples from different patients. Not randomizing methods within patients will cause patient effects to be confounded with treatment effects.

15.5. (a)

$$Y_{ij} = \mu + R_i + \tau_j + E_{ij} \qquad i = 1, 2, 3, 4; j = 1, 2, 3, 4, 5$$

where

Y_{ij} = distance the ball would travel on the jth cultivar from the ith slope,
μ = overall mean distance,
R_i = random effect of the ith slope, R_i are independent $N(0, \sigma_R^2)$,
τ_j = fixed effect of the jth cultivar, $\sum \tau_j = 0$,
E_{ij} = random error in the distance of the ball from the ith slope on the jth cultivar. The E_{ij} are independent $N(0, \sigma^2)$ and are independent of the R_i.

(b) The ANOVA table, as paraphrased from SAS, is

Dependent Variable: DISTANCE

Source	DF	Sum of Squares	Mean Square	F Value	Pr > F
BLOCK	3	1.0503269	0.3501090	25.79	0.0001
VARIETY	4	0.4512682	0.1128170	8.31	0.0019
Error	12	0.1628918	0.0135743		
Corrected Total	19	1.6644869			

The hypothesis tests for Block (Slope) and Treatment (Variety) are

H_0: $\sigma_R^2 = 0$
H_1: $\sigma_R^2 > 0$
Test Statistic: $F_{\text{Block}} = 25.79$;
 p–value $= 0.0001$
Conclusion: With this small of a p–value there is sufficient evidence to say that a Slope effect does exist, i.e. that there is variability in the distances that can be attributed to the slope.

H_0: $\psi_R^2 = 0$
H_1: $\psi_R^2 \neq 0$
Test Statistic: $F_{\text{Variety}} = 8.31$;
 p–value $= 0.0019$
Conclusion: Again, this p–value is less than most reasonable α–levels. We conclude that a significant difference in the mean distance of the varieties does exist.

(c) The LSD value for the Duncan method is $\text{LSD}_{ij}(\text{D}) = R(p, \nu, \alpha)\sqrt{\frac{MS[E]}{n}}$. For this data, $MS[E] = 0.013574$ and $n = 4$. There are five treatments, so we will need the .05–level Duncan's studentized range for 12 degrees of freedom and values of p at 2, 3, 4, and 5. We can look these up in Table C.12 in the appendix to get the following LSD values

p	2	3	4	5
$R(p, \nu, \alpha)$	3.081	3.225	3.313	3.370
$\text{LSD}_{ij}(\text{D})$	0.1795	0.1879	0.1930	0.1963

The sorted treatment means (coincidentally in the same order as the varieties are labeled) are

Variety	A	B	C	D	E
Mean	2.86350	2.68900	2.56700	2.54375	2.41950

As before, treatments that are connected by the horizontal line are not significantly different. For example, Varieties A and B are not statistically significantly different from one another, but Varieties A and C are.

15.7. (a) A two-factor RCB model with random Blocks and fixed factors can be written as

$$Y_{ijk} = \mu + R_i + \alpha_j + \beta_k + (\alpha\beta)_{jk} + E_{ijk} \qquad i = 1,\ldots,r; j = 1,\ldots,a; k = 1,\ldots,b.$$

The subscript for the block effect is i. The effects with a subscript of i are R_i and E_{ijk}. Therefore, the expected mean squares for blocks are a linear combination of σ_R^2 and σ^2. The remaining coefficients for the first effect are j and k to give the coefficient ab. There are no remaining coefficients for E_{ijk} so its coefficient is one. This gives us

$$EMS[R] = ab\sigma_R^2 + \sigma^2.$$

Using the same rules, the rest of the expected mean squares are

$$EMS[A] = rb\psi_A^2 + \sigma^2,$$
$$EMS[B] = ra\psi_B^2 + \sigma^2,$$
$$EMS[AB] = r\psi_{AB}^2 + \sigma^2, \quad \text{and}$$
$$EMS[E] = \sigma^2.$$

For testing blocks and the two factors the appropriate F-ratio is the mean square for the respective effect divided by the mean square for error.

(b) For Blocks random, A fixed, and B random the model is

$$Y_{ijk} = \mu + R_i + \alpha_j + B_k + (\alpha B)_{jk} + E_{ijk} \qquad i = 1,\ldots,r; j = 1,\ldots,a; k = 1,\ldots,b.$$

The expected mean squares are

$$EMS[R] = ab\sigma_R^2 + \sigma^2,$$
$$EMS[A] = rb\psi_A^2 + r\sigma_{\alpha B}^2 + \sigma^2,$$
$$EMS[B] = ra\sigma_B^2 + r\sigma_{\alpha B}^2 + \sigma^2,$$
$$EMS[AB] = r\sigma_{\alpha B}^2 + \sigma^2, \quad \text{and}$$
$$EMS[E] = \sigma^2.$$

From this, we have the following F-ratios

$$H_0\colon \sigma_R^2 = 0 \quad \text{is tested by} \quad F = \frac{MS[R]}{MS[E]},$$

$$H_0\colon \psi_A^2 = 0 \quad \text{is tested by} \quad F = \frac{MS[A]}{MS[AB]},$$

$$H_0\colon \sigma_B^2 = 0 \quad \text{is tested by} \quad F = \frac{MS[B]}{MS[AB]},$$

$$H_0\colon \sigma_{AB}^2 = 0 \quad \text{is tested by} \quad F = \frac{MS[AB]}{MS[E]},$$

(c) For all effects random the model becomes

$$Y_{ijk} = \mu + R_i + A_j + B_k + (AB)_{jk} + E_{i(jk)} \qquad i = 1, \ldots, r; j = 1, \ldots, a; k = 1, \ldots, b.$$

The expected mean squares are

$$EMS[R] = ab\sigma_R^2 + \sigma^2,$$

$$EMS[A] = rb\sigma_A^2 + r\sigma_{AB}^2 + \sigma^2,$$

$$EMS[B] = ra\sigma_B^2 + r\sigma_{AB}^2 + \sigma^2,$$

$$EMS[AB] = r\sigma_{AB}^2 + \sigma^2, \quad \text{and}$$

$$EMS[E] = \sigma^2.$$

From this table, we have the following F-ratios

$$H_0\colon \sigma_R^2 = 0 \quad \text{is tested by} \quad F = \frac{MS[R]}{MS[E]},$$

$$H_0\colon \sigma_A^2 = 0 \quad \text{is tested by} \quad F = \frac{MS[A]}{MS[AB]},$$

$$H_0\colon \sigma_B^2 = 0 \quad \text{is tested by} \quad F = \frac{MS[B]}{MS[AB]},$$

$$H_0\colon \sigma_{AB}^2 = 0 \quad \text{is tested by} \quad F = \frac{MS[AB]}{MS[E]},$$

15.9. (a) The pen is the experimental unit, so the sample size is twelve. There are three pens for each treatment. The treatment and main effect totals are

Barley Processing

N-Suppl.	Whole	Steamrolled	
Canola	23.1	21.6	44.7
Urea	21.3	20.1	41.4
	44.4	41.7	86.1

$$SS[Trt] = \frac{Y_{11+}^2}{n_{11+}} + \frac{Y_{12+}^2}{n_{12+}} + \frac{Y_{21+}^2}{n_{21+}} + \frac{Y_{22+}^2}{n_{22+}} - \frac{Y_{+++}^2}{N}$$

$$= \frac{23.1^2}{3} + \frac{21.6^2}{3} + \frac{21.3^2}{3} + \frac{20.1^2}{3} - \frac{86.1^2}{12} = 619.29 - 617.77 = 1.5225$$

$$SS[A] = \frac{Y_{1++}^2}{n_{1++}} + \frac{Y_{2++}^2}{n_{2++}} - \frac{Y_{+++}^2}{N} = \frac{44.7^2}{6} + \frac{41.4^2}{6} - \frac{86.1^2}{12}$$

$$= 618.675 - 617.768 = 0.907$$

$$SS[B] = \frac{Y_{+1+}^2}{n_{+1+} + \frac{Y_{+2+}^2}{n_{+2+}} - \frac{Y_{+++}^2}{N}} = \frac{44.4^2}{6} + \frac{41.7^2}{24} - \frac{86.1^2}{12}$$

$$= 618.375 - 617.768 = 0.607$$

$$SS[AB] = SS[Trt] - SS[A] - SS[B] = 1.5225 - 0.907 - 0.607 = 0.0085$$

Each of these, except $SS[Trt]$ has one degree of freedom. There are two degrees of freedom for blocks and forty-seven total degrees of freedom. Thus, there are forty-two degrees of freedom for error. With $MS[E] = 0.0588$, this gives the following hypothesis tests.

H$_0$: $\psi_{AB}^2 = 0$

H$_1$: $\psi^2_{AB} \neq 0$

Test Statistic: $F_c = 0.0085/0.0588 = 0.1446$

Rejection Region: Reject H$_0$ if $F_c > F(1, 6, .05)5.987$

Conclusion: Do not Reject H$_0$. The data does not support the conclusion that a significant interaction exists.

Since we did not reject the test for interaction, we now test the main effects.

H$_0$: $\psi^2_A = 0$

H$_1$: $\psi^2_A \neq 0$

Test Statistic:
$$F_c = 0.907/0.0588 = 15.42$$

Rejection Region: Reject H$_0$ if
$$F_c > F(1, 6, .05)5.987$$

Conclusion: Reject H$_0$. The data does show, at the 0.05 significance level, that a difference in the mean dry matter intake for the two N-supplements exists.

H$_0$: $\psi^2_B = 0$

H$_1$: $\psi^2_B \neq 0$

Test Statistic:
$$F_c = 0.607/0.0588 = 10.3231$$

Rejection Region: Reject H$_0$ if
$$F_c > F(1, 6, .05)5.987$$

Conclusion: Reject H$_0$. The data does show, at the 0.05 significance level, that a difference in the mean Dry Matter Intake for the two Barley Processing methods does exist.

(b) Because of our conclusions in Part a. we will calculate confidence intervals for the difference in the main effect means. The standard error for this difference is $\sqrt{2\frac{MS[E]}{rbn}}$ $= \sqrt{2\frac{0.0588}{12}} = 0.0990$. The t-statistic is $t(6, 0.05) = 1.9432$. Then the first confidence interval is

$$\overline{Y}_{1++} - \overline{Y}_{2++} \pm (1.9432)(0.0990) = 0.55 \pm 0.1924 = (0.3576, 0.7424)$$

so that with 95% confidence we can say that the mean Dry Matter Intake for Canola Meal is between 0.3576 and 0.7424 kg/day greater than the mean for Urea.
The second confidence interval is

$$\overline{Y}_{+1+} - \overline{Y}_{+2+} \pm (1.9432)(0.0990) = 0.45 \pm 0.1924 = (0.2576, 0.6424)$$

so that with 95% confidence we can say that the mean Dry Matter Intake for the Whole method of processing barley is between 0.2576 and 0.6424 kg/day greater than the mean for the Steamrolled method. Of course, while these differences are statistically significant, it is up to the researcher to determine if they are practically significant.

15.11. (a) The model can be written as

$$Y_{ijk} = \mu + R_i + \alpha_j + \beta_k + (\alpha\beta)_{jk} + E_{ijk} \qquad i = 1, 2, 3; j = 1, 2; k = 1, 2, 3;$$

where

Y_{ijk} = the measured percent soil water content in Block i with bed height j and row spacing k,

μ = mean percent soil water content,

R_i = the random effect of the ith block; the R_i are independent $N(0, \sigma^2_R)$,

α_j = the fixed effect of the jth level of height of bed; the α_j satisfy the constraint $\sum \alpha_j = 0$,

β_k = the fixed effect of the kth row spacing; the β_k satisfy the constraint $\sum \beta_k = 0$,

$(\alpha\beta)_{jk}$ = the joint effect of the jth level of height of bed and the kth level of row spacing; the $(\alpha\beta)_{jk}$ satisfy the constraints $\sum_j (\alpha\beta)_{jk} = 0$ and $\sum_k (\alpha\beta)_{jk} = 0$,

E_{ijk} = experimental error; the E_{ijk} are independent $N(0, \sigma^2)$.

(b) The ANOVA table below is paraphrased from SAS.

Dependent Variable: MOISTURE

Source	DF	Sum of Squares	Mean Square	F Value	Pr > F
BLOCK	2	236.89333	118.44667	26.57	0.0001
HEIGHT	1	14.40056	14.40056	3.23	0.1025
SPACING	2	127.85333	63.92667	14.34	0.0012
HEIGHT*SPACING	2	9.23111	4.61556	1.04	0.3903
Error	10	44.58667	4.45867		
Corrected Total	17	432.96500			

(c) The estimate of the difference between the expected moisture contents at "low" and "high" bed heights is

$$\hat{\theta} = \overline{Y}_{+1+} - \overline{Y}_{+2} = 19.8778 - 18.0889 = 1.7889$$

The standard error for this estimate is

$$\hat{\sigma}_{\hat{\theta}} = \sqrt{\left(\frac{1}{9} + \frac{1}{9}\right) 4.45867} = 0.9954$$

Then, with $t(10, 0.025) = 2.228$, the 95% confidence interval is

$$1.7889 \pm 2.228(0.9954) = (-0.4288, 4.0066)$$

With 95% confidence we conclude that the difference between the expected moisture contents at "low" and "high" bed heights is between -0.4288 and 4.0066 percent.

(d) The LSD value for the Duncan method is

$$LSD_{ij}(D) = R(p, \nu, \alpha)\sqrt{\frac{2MS[E]}{n}} = R(p, 10, 0.05)\sqrt{\frac{2(4.4587)}{6}}$$

Then for the values of p this is

p	$R(p, 10, 0.05)$	$LSD_{ij}(D)$
2	2.2278	2.716
3	2.3287	2.839

This gives the following table of results

Spacing	0.25	0.5	1.0
Mean	22.75	17.217	16.983

We conclude that Spacings 0.5 and 1.0 are not statistically significantly different, but that Spacing 0.25 is significantly different from the other two.

15.13. (a) To do simultaneous tests for the three hypotheses we need point estimates and standard errors. The point estimates can be found by substituting in the sample means as follows:

$$\hat{\theta}_{12} = \frac{6.11 + 4.59 + 3.92 + 3.52}{4} - \frac{3.10 + 2.98 + 4.60 + 4.82}{4} = 4.535 - 3.875 = .66,$$

$$\hat{\theta}_{13} = 4.535 - \frac{4.21 + 4.29 + 4.20 + 5.22}{4} = 4.535 - 4.48 = .055,$$

$$\hat{\theta}_{14} = 4.535 - \frac{9.01 + 8.30 + 6.53 + 8.46}{4} = 4.535 - 8.075 = -3.54.$$

The estimated standard error for all three of the contrasts has the same general form

$$\hat{\sigma}_{\hat{\theta}_{1j}} = \sqrt{\frac{1^2 + (-1)^2}{4} MS[E]} = \sqrt{\frac{0.7347}{2}} = 0.6061.$$

To do a simultaneous test of the three null hypotheses, we can use the Bonferroni approach and test each hypothesis at the α level $\frac{\alpha}{3}$.

For an $\alpha = 0.05$ level test we compare $\hat{\theta}_{ij}$ to $t(6, 3)0.05\hat{\sigma}_{\hat{\theta}_{ij}} = (3.287)(0.6061) = 1.99$.

The comparisons are

$$|\hat{\theta}_{12}| = 0.66 < 1.99$$
$$|\hat{\theta}_{13}| = 0.06 < 1.99$$
$$|\hat{\theta}_{14}| = 5.841 > 1.99$$

Only H_{014} is rejected. At the $\alpha = 0.05$ level we conclude that of all comparisons with treatment 1 simultaneously only treatment 4 differs from treatment 1.

(b) Using the above calculations and the Bonferroni method we get that each interval has the form

$$\hat{\theta}_{1j} \pm t(6, 0.0083)\hat{\sigma}_{\hat{\theta}_{ij}}.$$

This gives the following three intervals

$$
\begin{array}{lll}
0.66 \pm 3.287(0.6061) & \text{or} & (-1.3345, 2.6545), \\
0.055 \pm 3.287(0.6061) & \text{or} & (-1.9395, 2.0495), \\
-3.54 \pm 3.287(0.6061) & \text{or} & (-5.5345, -1.5455).
\end{array}
$$

With 95% confidence we conclude that the true values of the three contrasts are within their respective intervals.

(c) The Tukey test compares the absolute value of each paired difference to the value

$$q(4, 6, 0.05)\sqrt{\frac{0.7347}{4}} = 4.90(0.4286) = 2.1.$$

All of the sample means have already been calculated as 4.535, 3.875, 4.480 and 8.075 for treatments 1 through 4, respectively. On making the comparison it is found that the only difference is that treatment 4 is significantly different(higher) than all other treatments.

(d) The analysis in Part Part b in Exercise **15.13** is most appropriate among the three because in this analysis, we estimate ranges of plausible values for the differences of interest.

(e) Because the hypothesis tests for θ_{12} and θ_{13} did not reject the null hypothesis we can not say that these methods produce results that is not comparable to the results from the standard method. Because the hypothesis test for θ_{14} did reject the null hypothesis we can say that this method is not comparable to the standard method. These statements can be made at the .05 significance level. We must be careful concluding that Methods 2 and 3 are comparable to the standard method, because we run the risk of making a Type II Error and we do not know what that risk is, i.e. we do not know the probability of a Type II Error.

15.15. (a) Let Y_{ijkl} represent the cats tolerance in (10log(dose)) measured by the ith observer at time j (j=1(AM), j=2(PM)) on the kth day given the lth Cardiac Substance. Then

$$
\begin{aligned}
Y_{ijkl} &= \mu + R_i + \pi_j + (R\pi)_{ij} + C_k + \tau_l + E_{ijkl} \\
&\quad i = 1, 2; j = 1, 2; k = 1, \ldots, 4; l = 1, \ldots, 4;
\end{aligned}
$$

where

$$\mu \quad = \quad \text{the overall mean tolerance,}$$

$$R_i \quad = \quad \text{the random effect of the } i^{th} \text{ observer, the } R_i \text{ are ind. } N(0, \sigma_R^2),$$

$$\pi_j \quad = \quad \text{the fixed effect of the } j^{th} \text{ time, } \sum \pi_j = 0,$$

$$(R\pi)_{ij} \quad = \quad \text{the random joint effect of the } j^{th} \text{ time and the } i^{th} \text{ observer,}$$
$$\text{the } (R\pi)_{ij} \text{ are ind. } N(0, \sigma_{R\pi}^2),$$

$$C_k \quad = \quad \text{the random effect of the } k^{th} \text{ day, the } C_k \text{ are ind. } N(0, \sigma_C^2),$$
$$\text{the } C_k, (R\pi)_{ij} \text{ and } R_i \text{ are mutually independent,}$$

$$\tau_l \quad = \quad \text{the fixed effect of the } l^{th} \text{ Cardiac Substance, } \sum \tau_l = 0,$$

$$E_{ijkl} \quad = \quad \text{the random error, } E_{ijkl} \text{ are ind. } N(0, \sigma^2),$$
$$\text{the } E_{ijkl} \text{ are independent of } C_k, (R\pi)_{ij} \text{ and } R_i.$$

(b) The following ANOVA table was generated using Proc ANOVA is SAS.

Dependent Variable: TOL

Source	DF	Sum of Squares	Mean Square	F Value	Pr > F
Model	9	1.7962563	0.1995840	0.66	0.7228
Error	6	1.8092875	0.3015479		
Corrected Total	15	3.6055438			

R-Square	C.V.	Root MSE	TOL Mean
0.498193	15.99516	0.5491	3.4331

Source	DF	Anova SS	Mean Square	F Value	Pr > F
OBS	1	0.0715563	0.0715563	0.24	0.6434
TIME	1	0.0297563	0.0297563	0.10	0.7641
OBS*TIME	1	0.0052562	0.0052562	0.02	0.8993
DAY	3	1.4274688	0.4758229	1.58	0.2900
SUBSTANCE	3	0.2622188	0.0874063	0.29	0.8315

(c) Based on the above test for TREAT, there is no evidence($p = .8315$) to conclude that differences exist among the four Cardiac Substances.

(d) The sample means can be calculated as

$$\overline{Y}_A = 3.5275 \quad \overline{Y}_B = 3.235 \quad \overline{Y}_C = 3.5625 \quad \overline{Y}_D = 3.4075.$$

Using Tukey's test we compare in absolute value each paired difference to the value

$$q(4, 6, .05)\sqrt{\frac{MS[E]}{4}} = 4.90\sqrt{\frac{.3015}{4}} = 1.3453.$$

In making the comparisons it is found that none of the Cardiac Substances are found to be different from one another.

15.17. (a) The following graphs can be used to check for the assumptions of normality, additivity and equal variances.

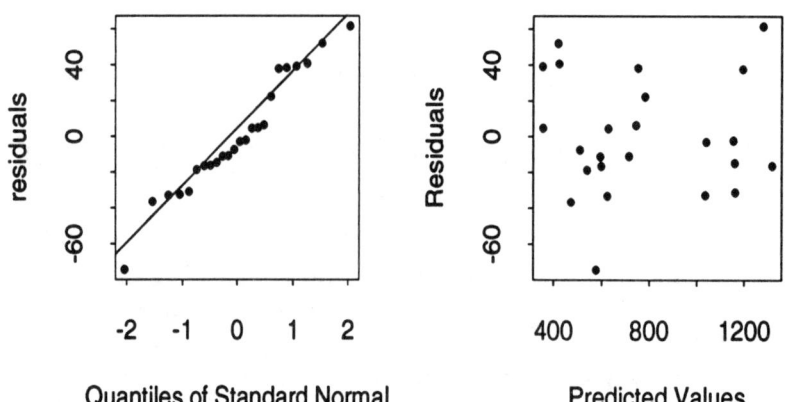

Normal Q-Q Plot

residuals

Quantiles of Standard Normal

Residuals vs. Predicted

Residuals

Predicted Values

Neither of these shows any problems with the assumptions.

(b) The following table shows the pieces necessary to calculate $SS[NA]$ where $Z_{ij} = (\overline{Y}_{i+} - \overline{Y}_{++})(\overline{Y}_{+j} - \overline{Y}_{++})$.

Obs.	Patient (i)	Method (j)	Y_{ij}	$\overline{Y}_{i+} - \overline{Y}_{++}$	$\overline{Y}_{+j} - \overline{Y}_{++}$	Z_{ij}
1	1	1	360	−345.1667	−66.8333	23068.639
2	1	2	435	−345.1667	49.5000	−17085.750
3	1	3	391	−345.1667	−70.1667	24219.194
4	1	4	502	−345.1667	87.5000	−30202.083
5	2	1	1035	337.5833	−66.8333	−22561.819
6	2	2	1152	337.5833	49.5000	16710.375
7	2	3	1002	337.5833	−70.1667	−23687.097
8	2	4	1230	337.5833	87.5000	29538.542
9	3	1	632	−72.9167	−66.8333	4873.264
10	3	2	750	−72.9167	49.5000	−3609.375
11	3	3	591	−72.9167	−70.1667	5116.319
12	3	4	804	−72.9167	87.5000	−6380.208
13	4	1	581	−102.9167	−66.8333	6878.264
14	4	2	703	−102.9167	49.5000	−5094.375
15	4	3	583	−102.9167	−70.1667	7221.319
16	4	4	790	−102.9167	87.5000	−9005.208
17	5	1	463	−278.1667	−66.8333	18590.806
18	5	2	520	−278.1667	49.5000	−13769.250
19	5	3	471	−278.1667	−70.1667	19518.028
20	5	4	502	−278.1667	87.5000	−24339.583
21	6	1	1131	461.5833	−66.8333	−30849.153
22	6	2	1340	461.5833	49.5000	22848.375
23	6	3	1144	461.5833	−70.1667	−32387.764
24	6	4	1300	461.5833	87.5000	40388.542

$SS[NA]$ can now be calculated as

$$SS[NA] = \frac{\left(\sum_{i=1}^{r} \sum_{j=1}^{t} Y_{ij} Z_{ij} \right)^2}{\sum_{i=1}^{r} \sum_{j=1}^{t} Z_{ij}^2} = \frac{9057227^2}{10517360029} = 7799.805$$

From Exercise **15.3** the $SS[E] = 24593.7$ with 15 degrees of freedom. Hence,

$$MSE[NA] = \frac{SS[E] - SS[NA]}{df_E - 1} = \frac{24593.7 - 7799.805}{15 - 1} = 1199.5639.$$

Then the test statistic is

$$F_c = \frac{SS[NA]}{MSE[NA]} = \frac{7799.805}{1199.5639} = 6.5022.$$

We reject H_{0NA}: the additive model is appropriate at the 0.05 significance level when $F_c > F(1, 14, 0.05) = 4.60$. Therefore, at the 0.05 significance level, the data does not show evidence of significant non-additivity.

15.19. (a) The following plots will help check for assumption violations.

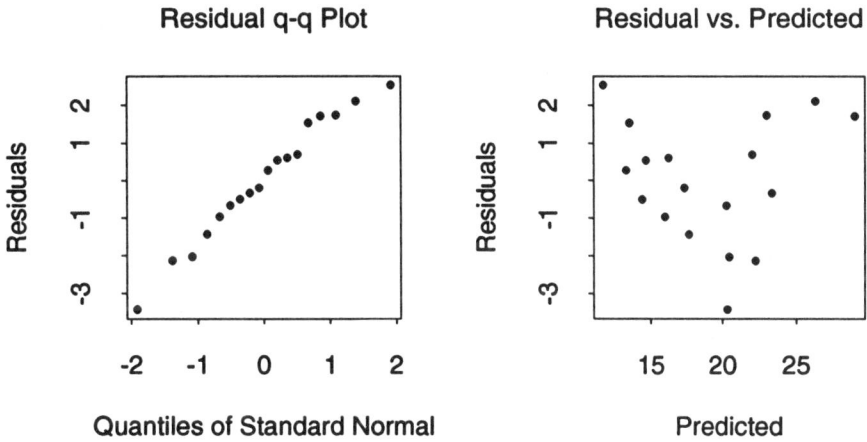

(b) To apply Tukey's test to the data set we will complete the table as in the text.

i	j	Y_{ij}	$\overline{Y}_{i+} - \overline{Y}_{++}$	$\overline{Y}_{+j} - \overline{Y}_{++}$	Z_{ij}	$Y_{ij}Z_{ij}$
1	1	30.7	5.1167	4.9000	25.0718	769.70426
1	2	24.7	5.1167	−1.1333	−5.7988	−143.23036
1	3	16.9	5.1167	−3.7667	−19.2731	−325.71539
2	1	22.7	−1.8833	4.9000	−9.2282	−209.48014
2	2	15.0	−1.8833	−1.1333	2.1343	32.01450
2	3	13.6	−1.8833	−3.7667	7.0938	96.47568
3	1	23.0	−.5533	4.9000	−2.7111	−62.35530
3	2	17.1	−.5533	−1.1333	.6271	10.72341
3	3	15.2	−.5533	−3.7667	2.0841	31.67832
4	1	28.4	2.4167	4.9000	11.8418	336.30712
4	2	19.6	2.4167	−1.1333	−2.7388	−53.68048
4	3	16.2	2.4167	−3.7667	−9.1030	−147.46860
5	1	20.1	−1.6500	4.9000	−8.085	−162.50850
5	2	16.8	−1.6500	−1.1333	1.8699	31.41432
5	3	15.1	−1.6500	−3.7667	6.2151	93.84801
6	1	18.4	−3.4500	4.9000	−16.905	−311.05200
6	2	13.9	−3.4500	−1.1333	3.9089	54.33371
6	3	14.3	−3.4500	−3.7667	12.9951	185.82993

Hence,

$$\sum_{i=1}^{6}\sum_{j=1}^{3} Z_{ij}^2 = 1993.814 \qquad \sum_{i=1}^{6}\sum_{j=1}^{3} Y_{ij}Z_{ij} = 226.8385,$$

and

$$SS[NA] = \frac{226.8385^2}{1993.814} = 25.8077.$$

The denominator of the test statistic is

$$MS[NA] = \frac{SS[E] - SS[NA]}{10 - 1} = \frac{44.59 - 25.8077}{9} = 2.347788.$$

The calculated value of the test statistic is

$$F_c = \frac{25.8077}{2.3478} = 10.9923,$$

which is compared to an $F(1,9)$. The associated p−value is 0.0090. Since this p−value is smaller than almost all α levels used in practice, we would reject the null hypothesis and conclude that the additive model is insufficient to fully explain what is going on. This can be further seen in the pattern of residuals. There seems to be interaction amongst the treatments and blocks.

15.21. (a) The following plots can be used to check for violation of assumptions.

Normal Q-Q Plot

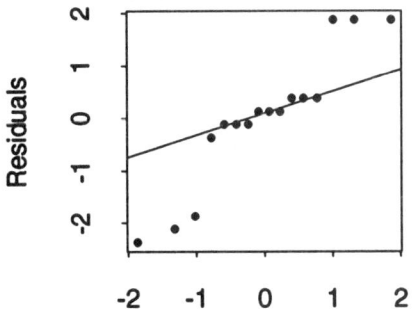

Quantiles of Standard Normal

Residuals vs. Predicted

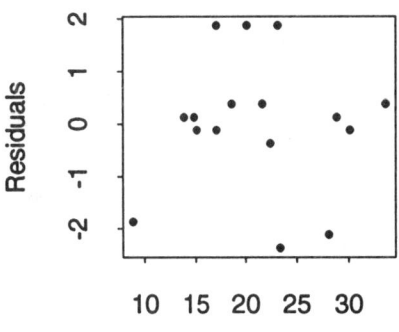

Predicted Values

There might be some problems with the normality and/or the appropriateness of an additivity model.

(b) Using the data in Exercise **15.14** and the methods in the text the following table can be created to facilitate calculating SS[NA] where

$$Z_{ij} = (\overline{Y}_{i++} - \overline{Y}_{+++})(\overline{Y}_{+j+} - \overline{Y}_{+++}) + (\overline{Y}_{i++} - \overline{Y}_{+++})(\overline{Y}_{++k} - \overline{Y}_{+++})$$
$$+ (\overline{Y}_{+j+} - \overline{Y}_{+++})(\overline{Y}_{++k} - \overline{Y}_{+++}).$$

i	j	k	Y_{ijk}	$\overline{Y}_{i++} - \overline{Y}_{+++}$	$\overline{Y}_{+j+} - \overline{Y}_{+++}$	$\overline{Y}_{++k} - \overline{Y}_{+++}$	Z_{ijk}
1	1	AH	14	-5.3125	0.9375	-2.8125	7.3242
1	2	AL	30	0.4375	0.9375	7.6875	10.9805
1	3	BH	15	-0.5625	0.9375	-6.3125	-2.8945
1	4	BL	29	5.4375	0.9375	1.4375	14.2617
2	1	AL	25	-5.3125	-0.3125	7.6875	-41.5820
2	2	BH	15	0.4375	-0.3125	-6.3125	-0.9258
2	3	BL	22	-0.5625	-0.3125	1.4375	-1.0820
2	4	AH	21	5.4375	-0.3125	-2.8125	-16.1133
3	1	BL	17	-5.3125	-0.0625	1.4375	-7.3945
3	2	AH	19	0.4375	-0.0625	-2.8125	-1.0820
3	3	AL	26	-0.5625	-0.0625	7.6875	-4.7695
3	4	BH	22	5.4375	-0.0625	-6.3125	-34.2695
4	1	BH	7	-5.3125	-0.5625	-6.3125	40.0742
4	2	BL	22	0.4375	-0.5625	1.4375	-0.4258
4	3	AH	19	-0.5625	-0.5625	-2.8125	3.4805
4	4	AL	34	5.4375	-0.5625	7.6875	34.4180

(Note that Order is i, Patients is j, and Treatment is k.) The Sum of Squares for Non-additivity is

$$SS[NA] = \frac{\left(\sum_{i=1}^{t} \sum_{j=1}^{t} \sum_{k=1}^{t} Y_{ijk} Z_{ijk} \right)^2}{\sum_{i=1}^{t} \sum_{j=1}^{t} \sum_{k=1}^{t} Z_{ijk}^2} = \frac{-130.207^2}{6432.554} = 2.6356$$

Note that the summations are done over those values of i, j, and k for which Y_{ijk} is defined. Next, the Mean Square is (from Exercise **15.14** $SS[E] = 24.8750$ with six degrees of freedom)

$$MSE[NA] = \frac{SS[E] - SS[NA]}{df_E - 1} = \frac{24.8750 - 2.6356}{6 - 1} = 4.4479$$

The test statistic is

$$F_c = \frac{SS[NA]}{MSE[NA]} = \frac{2.6356}{4.4479} = 0.5925$$

We reject H_{ONA}: the additive model is appropriate at the 0.05 significance level when $F_c > F(1, 5, 0.05) = 6.61$. Therefore, we do not reject H_{ONA}.

15.23. (a) A model under which the Friedman test can be used to compare the four methods is

$$Y_{ij} = \mu + R_i + \tau_j + E_{ij} \qquad i = 1, 2, 3, \ldots, 6; j = 1, 2, 3, 4,$$

where

Y_{ij} = the measured serum amylase value for the ith patient with the jth method,

μ = overall mean serum amylase value,

R_i = the random effect of the ith patient; the R_i constitute a random sample from a population of patients,

τ_j = the fixed effect of the jth method; $\sum \tau_j = 0$,

E_{ij} = random error. The E_{ij} are independent and have a common continuous distribution. The E_{ij} and R_i are independent.

(b) The table of ranks of the observations is

	Method			
Patient	1	2	3	4
1	1	3	2	4
2	2	3	1	4
3	2	3	1	4
4	1	3	2	4
5	1	4	2	3
6	1	4	2	3
Sum	8	20	10	22

The test statistic is calculated as

$$W_1 = 12 \sum_{j=1}^{t} R_{+j}^2 - 3r^2 t(t+1)^2 = 12(8^2 + 20^2 + 10^2 + 22^2) - 3(6)^2(4)(5)^2$$

$$= 12576.0 - 10800.0 = 1776.0,$$

$$C = rt(t+1) - \frac{1}{t-1} \sum_{i=1}^{r} \sum_{k=1}^{g_i} t_{ik}(t_{ik}^2 - 1) = 6(4)(5) - 0 = 120$$

(since there are no ties), and

$$W = \frac{W_1}{C} = 14.8$$

The Friedman test is

H_0: $\tau_1 = \tau_2 = \tau_3 = \tau_4$

H_1: $\tau_j \neq \tau_{j'}$ for at least one pair $j \neq j'$

Test Statistic: $W_c = 14.8$

Rejection Region: $Reject$H$_0$ if W $>$ $\chi^2(3, 0.01) = 11.3449$

Conclusion: At the 0.01 significance level H$_0$ is rejected. There is sufficient evidence to say that there is a difference between the four Methods.

(c) In Exercise **15.3**, the conclusion was to reject H$_0$ which is consistent with the conclusion in part a. Because the Friedman Rank Sum test required less assumptions, we feel more confident in our conclusions; especially, if there was some question about the validity of the ANOVA assumptions.

15.25. (a) To use the Friedman test we need to rank within each Donor. The table with the observed values replaced by their ranks follows.

Donor	1	2	3	4	5	6	7	8	Total
Hanks	1	2	1.5	2	2	1	1	2	12.5
Sucrose	2	1	1.5	1	1	2	2	1	11.5

$$W_1 = 12[12.5^2 + 11.5^2] - 3(8)^2 2(3)^2 = 3462 - 3456 = 6,$$

and there is one group with two ties. All of the other groups have no ties, hence

$$C = 8(2)3 - \frac{1}{1}2(2^2 - 1) = 48 - 6 = 42.$$

The test statistic has the form

$$W_c = \frac{6}{42} = 0.1429.$$

At the 0.05 significance level, there is not enough evidence to say that the two solutions lead to different results.

(b) The same conclusion is made with both procedures. The two procedures lead to similar conclusions. Either procedure could be used to reach the same conclusion in this setting.

(c) The model would have the following form

$$Y_{ij} = \mu + R_i + \tau_j + E_{ij} \quad i = 1, \ldots, 8; j = 1, 2;$$

where

$$\mu = \text{the overall mean},$$
$$R_i = \text{the Random Effect of the } i^{th} \text{ Donor},$$
$$\tau_j = \text{the fixed effect of the } j^{th} \text{ Solution},$$
$$E_{ij} = \text{the random error}.$$

The model appropriate for the sign test is that

$$Y_{i1} - Y_{i2} = \mu + E_i,$$

where

$$\mu = \text{the mean difference which does not depend on } i,$$
$$E_i = \text{the random error}.$$

This is what one gets from the above model when looking at the difference. Thus, the assumption for the Friedman test implies the assumption for the sign test, but the converse is not true.

15.27. Using the information in the text we can reconstruct the table to look like the one in the text. Since thirteen specimens grew in all three media, there will be thirteen rows with all ones. Since two grew in B and C, there will be two rows with a zero in the A column and ones in the B and C column. In this case i represents reps and j represents treatments.

Reps	A	B	C	Total
1	1	1	1	
2	1	1	1	
$\vdots$	$\vdots$	$\vdots$	$\vdots$	
13	1	1	1	13
14	1	1	0	
$\vdots$	$\vdots$	$\vdots$	$\vdots$	
19	1	1	0	6
20	1	0	1	
21	1	0	1	
22	1	0	1	
23	1	0	1	4
24	0	1	1	
25	0	1	1	2
26	0	0	0	
$\vdots$	$\vdots$	$\vdots$	$\vdots$	
100	0	0	0	75
Total	23	21	19	63

The test statistic Q_c can now be calculated.

$$
\begin{aligned}
Q_c &= \frac{3(3-1)[23^2 + 21^2 + 19^2] - (3-1)63^2}{3(63) - [13(3^2) + 12(2)^2]} \\
&= \frac{7986 - 7938}{189 - 165} = \frac{48}{24} = 2
\end{aligned}
$$

This value is compared to a $\chi^2(2)$, and has associated p-value of 0.3679. Based on this value we do not reject the null hypothesis; there is not enough evidence to suggest that the bacteria grow differently in any of the three media. The only assumption necessary is that the 3 parts of the bacteria within each specimen were randomly assigned to one of the three growth media.

Chapter 16

Repeated Measures Studies

16.1. (a) The reason that this study must be regarded as a repeated measures study is that all measurements taken for an individual subject will most likely be correlated with one another. It would seem unreasonable that the observations for the same subject in different periods are independent. Specifically it would be more natural that measurements in successive periods for the same subject to be highly correlated.

(b) The **between units factor** is Treatment with two levels (Drug, Placebo). The **within units factor** is the Period with 6 levels (1, 2, 3, 4, 5, and 6).

(c) Main–units in this study correspond to the subjects and sub–units in this study correspond to the period measurements within subject.

16.3. (a) The expected cholesterol level at Period j for a subject receiving Treatment i, μ_{ij}, could be written as

$$\mu_{ij} = \mu + \alpha_i + \beta_j + (\alpha\beta)_{ij},$$

where

$$
\begin{aligned}
\mu &= \quad \text{overall mean cholesterol level} \\
\alpha_i &= \quad \text{the effect of the } i^{th} \text{ Treatment on mean cholesterol level} \\
\beta_j &= \quad \text{the effect of the } j^{th} \text{ Period on mean cholesterol level} \\
(\alpha\beta)_{ij} &= \quad \text{the interaction effect of the } i^{th} \text{ Treatment effect during the } j^{th} \text{ Period on} \\
&\qquad \text{mean cholesterol level.}
\end{aligned}
$$

(b)
$$Y_{ijk_i} = \mu_{ij} + S_{ijk_i} + E_{ijk_i} \quad i = 1,2; j = 1,\ldots,6; k_1 = 1,\ldots,12, k_2 = 1,\ldots,11;$$

where

Y_{ijk_i} = the measured cholesterol level at the jth Period for the k_i^{th} subject receiving the ith Treatment

μ_{ij} = the expected cholesterol level at Period j for a subject receiving Treatment i

S_{ijk_i} = a random quantity that equals the difference between μ_{ij} and the **expected** cholesterol level during the jth Period for the k_i^{th} subject receiving the ith Treatment

E_{ijk_i} = a random quantity that equals the amount by which the cholesterol level during the jth Period of the k_i^{th} subject receiving the ith Treatment differs from the subject's expected cholesterol level measurement

(c) No, it may not be reasonable that the correlation between Period measurements within a subject are the same. We would expect measurements taken close together in time to be correlated more highly than measurements taken further apart. In other words, measurements taken at Periods 1 and 2 for a subject should be more highly correlated than measurements taken at Periods 1 and 6 for a subject.

16.5. Writing Y_{ij} as in the text gives

$$
\begin{aligned}
\text{Var}(Y_{ij}) &= \text{Var}(\mu_{ij} + S_{ij} + E_{ij}) \\
&= \text{Var}(S_{ij} + E_{ij}).
\end{aligned}
$$

Now by using the fact that S_{ij} and E_{ij} are independent normal random variables and the results of Chapter 9 we get that

$$
\begin{aligned}
\text{Var}(Y_{ij}) &= \text{Var}(S_{ij} + E_{ij}) \\
&= \text{Var}(S_{ij}) + \text{Var}(E_{ij}) \\
&= \sigma_{sj}^2 + \sigma_j^2.
\end{aligned}
$$

16.7. (a)

$$
Y_{ijk} = \mu + \alpha_i + \beta_j + (\alpha\beta)_{ij} + S_{ik} + E_{ijk}, \quad i = 1, 2, 3; , j = 1, \ldots, 4; k = 1, 2,
$$

where

$$
\begin{aligned}
\mu &= \text{the overall mean crop yield,} \\
\alpha_i &= \text{the fixed effect of the } i^{th} \text{ irrigation method,} \\
\beta_j &= \text{the fixed effect of the } j^{th} \text{ fertilization method,} \\
(\alpha\beta)_{ij} &= \text{the joint effect of the } i^{th} \text{ irrigation method and } j^{th} \text{ fertilization method,} \\
S_{ik} &= \text{the effect of the } k^{th} \text{ main--plot receiving the } i^{th} \text{ irrigation method,} \\
E_{ijk} &= \text{the error effect.}
\end{aligned}
$$

The usual assumptions about the parameters are also made as in the text.

(b) The completed ANOVA table follows.

Source	df	SS	MS	F_c	p-value
Irrig	2	287.0825	143.5413	95.0233	0.0019
Main-plot Error	3	4.5318	1.5106	——	——
Fert	3	114.1184	38.0395	56.2586	0.0001
Irrig*Fert	6	8.8662	1.4777	2.1855	0.1404
Sub-plot Error	9	6.0854	0.6762	——	——
Total	23	420.6843			

Based on the above F values, there is no evidence ($p = .1404$) of interaction between Irrigation method and Fertilizer method. However, both Fertilization method and Irrigation method are found to significantly ($p = 0.001$, and 0.0019) affect crop yield.

(c) Yes there does seem to be a significant difference between the irrigation methods. The appropriate standard error from the table can be calculated as

$$
\sqrt{\frac{2(1.5106)}{2(4)}} = 0.6145.
$$

Also it is easy to verify

$$
\bar{Y}_{1++} = 11.1338 \quad \bar{Y}_{2++} = 15.135 \quad \bar{Y}_{3++} = 6.6675.
$$

186

The Tukey method will be used to make the three comparisons. Each difference is compared to the value $q(3, 3, 0.05)(0.6145) = 5.91(0.6145) = 3.6317$. Making the comparisons, it can be concluded that all irrigation methods are significantly different from one another. (This is because all mean differences are larger than 3.6317, for example 15.135-11.1338=4.0012.)

(d) Yes there does seem to be a significant difference between the fertilization methods. The appropriate standard error from the table can be calculated as

$$\sqrt{\frac{2(.6762)}{2(3)}} = 0.4748.$$

Also it is easy to verify

$$\bar{Y}_{+1+} = 8.0367 \quad \bar{Y}_{+2+} = 9.7583 \quad \bar{Y}_{+3+} = 12.7317 \quad \bar{Y}_{+4+} = 13.385.$$

The Tukey method will once again be used to make the six comparisons. Each difference is compared to the value $q(4, 9, 0.05)(0.4748) = 4.41(0.4748) = 2.0939$. Upon making the comparisons it is found that the only significant differences are that Fertilization Methods 1 and 2 are different from Fertilization Methods 3 and 4. (These are the only sets whose differences are larger than the critical value 2.0939.)

(e) The three orthogonal contrasts defined as N, P and NP are as follows:

$$\hat{N} = (-1)\bar{Y}_{+1+} + (+1)\bar{Y}_{+2+} + (-1)\bar{Y}_{+3+} + (+1)\bar{Y}_{+4+}$$
$$\hat{P} = (-1)\bar{Y}_{+1+} + (-1)\bar{Y}_{+2+} + (+1)\bar{Y}_{+3+} + (+1)\bar{Y}_{+4+}$$
$$\hat{NP} = (+1)\bar{Y}_{+1+} + (-1)\bar{Y}_{+2+} + (-1)\bar{Y}_{+3+} + (+1)\bar{Y}_{+4+}$$

In the above equations N and P correspond to the contrasts associated with the main effects of nitrogen and phosphate respectively and NP corresponds to the contrast associated with the interaction effect. It is easy to verify that these contrasts are all pairwise orthogonal by noting that each contrasts coefficients are the numbers in parentheses. Proving N and P are orthogonal would amount to showing that the product of the coefficients for N with the corresponding coefficients for P sum to zero.

(f) A set of 95% Confidence intervals for the three contrasts are constructed using the Bonferroni method. The estimates for N, P and NP are found by replacing the population means with their sample means. This gives

$$\hat{N} = 2.3749 \quad \hat{P} = 8.3217 \quad \hat{NP} = -1.0683.$$

Each orthogonal contrast has a standard error of

$$\sqrt{\frac{4(0.6762)}{2(3)}} = 0.6714.$$

Hence a set of 95% Simultaneous Confidence bands are

$$\hat{N} \quad \pm \quad t(9, \tfrac{0.05}{6})(0.6714)$$
$$2.3749 \quad \pm \quad 1.9694 \quad \text{or} \quad (0.4055, 4.3443),$$
$$\hat{P} \quad \pm \quad 2.9333(0.6714)$$
$$8.3217 \quad \pm \quad 1.9694 \quad \text{or} \quad (6.3523, 10.2911),$$
$$\hat{NP} \quad \pm \quad 2.9333(0.6714)$$
$$1.0683 \quad \pm \quad 1.9694 \quad \text{or} \quad (-0.9011, 3.0377).$$

With 95% Confidence we conclude that simultaneously the three orthogonal contrasts all lie within their respective confidence interval.

(g) The following commands and their output from SAS find the necessary calculations for the ANOVA table

```
proc glm;
 classes irrig rep fert;
 model yield=irrig rep(irrig) fert irrig*fert;
 test h=irrig e=rep(irrig);
```

Dependent Variable: YIELD

Source	DF	Sum of Squares	Mean Square	F Value	Pr > F
Model	14	414.59887	29.61421	43.80	0.0001
Error	9	6.08539	0.67615		
Corrected Total	23	420.68426			

	R-Square	C.V.	Root MSE	YIELD Mean
	0.985535	7.489796	0.8223	10.979

Source	DF	Type III SS	Mean Square	F Value	Pr > F
IRRIG	2	287.08252	143.54126	212.29	0.0001
REP(IRRIG)	3	4.53176	1.51059	2.23	0.1536
FERT	3	114.11835	38.03945	56.26	0.0001
IRRIG*FERT	6	8.86624	1.47771	2.19	0.1404

Tests of Hypotheses using the Type III MS for
REP(IRRIG) as an error term

Source	DF	Type III SS	Mean Square	F Value	Pr > F
IRRIG	2	287.08252	143.54126	95.02	0.0019

16.9. (a)

$$Y_{ijk} = \mu + \alpha_i + \beta_j + (\alpha\beta)_{ij} + S_{ik} + E_{ijk},$$
$$i = 1,\ldots,4;, j = 1,\ldots,5; k = 1,2;$$

where

$\mu =$ the overall mean five year height,

$\alpha_i =$ the fixed effect of the ith Soil Type,

$\beta_j =$ the fixed effect of the jth Site Preparation Method,

$(\alpha\beta)_{ij} =$ the joint effect of the ith Soil Type and k^{th} Site Preparation Method,

$S_{ik} =$ the random effect of the kth Location having Soil Type i,

$E_{ijk} =$ the random error effect.

The usual assumptions about the parameters are also made as in the text.

(b) All but two of the SS's are found. The Main Error SS is simplest to calculate. To do so we need to add a vertical column to the table representing the total for the Soil Type and Location, represented by T_{ij}. The totals are

$$T_{11} = 47.361 \quad T_{12} = 46.171 \quad T_{21} = 69.209 \quad T_{22} = 49.790$$
$$T_{31} = 36.957 \quad T_{32} = 55.001 \quad T_{41} = 33.002 \quad T_{42} = 60.873.$$

The SS[Main-plot Error] can be found using the formula for computing SS[S(A)] as

$$SS[\text{Main-plot Error}] = \frac{1}{5}\sum T_{ij}^2 - \frac{1}{5(2)}\sum T_{i+}^2$$

$$= \frac{47.361^2 + \ldots + 60.873^2}{5} - \frac{93.532^2 + \ldots + 93.875^2}{10}$$

$$= \frac{20829.3593}{5} - \frac{40177.7864}{10} = 148.0893.$$

By subtraction

$$SS[\text{Sub-plot Error}] = SS[Tot] - \text{All other SS's} = 291.8767 - 265.5768 = 26.2999.$$

Using these results the completed ANOVA table has the following form.

Source	df	SS	MS	F_c	p-value
Soil Type	3	50.4317	16.8106	0.4541	0.7287
Main-Plot Error	4	148.0893	37.0223	——	——
Site Prep. Method	4	46.1080	11.5270	7.0128	0.0018
Soil Type×Site Prep.	12	20.9478	1.7457	1.0621	.4460
Sub-plot Error	16	26.2999	1.6437	——	——
Total	35	291.8767			

(c) Based on the ANOVA table, there is no evidence ($p = .4460$) of an interaction effect between Soil Type and Site Preparation method on mean yield. There is also no evidence ($p = 0.7289$) of a main effect due to Soil Type on mean yield. There is a significant ($p = 0.0018$) effect due to the main effect of Site Preparation Method on mean yield.

(d) Comparisons of the Soil Type correspond to comparisons of main effects, hence the appropriate standard error of the differences corresponds to

$$\widehat{\sigma}_{\bar{Y}_{i++} - \bar{Y}_{j++}} = \sqrt{\frac{2MS[\text{Main Error}]}{2(5)}} = \sqrt{\frac{2(37.0223)}{10}} = 2.7211.$$

The four Soil Type means can be calculated as

$$\bar{Y}_{1++} = 9.3532 \quad \bar{Y}_{2++} = 11.8999 \quad \bar{Y}_{3++} = 9.1958 \quad \bar{Y}_{4++} = 9.3875.$$

The Tukey method will be used to make the three comparisons between the comparisons with $\alpha = .05$. Each paired difference is compared to the value

$$q(4, 4)(2.7211) = 5.76(2.7211) = 15.6735.$$

None of the paired differences are larger than this value, hence (as expected) there is no evidence that any of the 4 soil types are different.

(e) An unbiased estimate of the contrast between the heights of plants for treated sites that are fertilized versus those that are not treated is

$$\begin{aligned} \widehat{\theta} &= \overline{Y}_{+1+} + \overline{Y}_{+2+} - \overline{Y}_{+3+} - \overline{Y}_{+4+} \\ &= 10.9881 + 11.4831 - 9.4288 - 8.7819 = 4.2605. \end{aligned}$$

The above contrast is a comparison of subplots, hence the appropriate standard error can be found as

$$\widehat{\sigma}_{\widehat{\theta}} = \sqrt{\frac{4MS[E]}{2(4)}} = \sqrt{\frac{4(1.6437)}{8}} = .9066.$$

The above comparison differs from that given in the table in the text since it is a comparison of four means rather than two. Hence, the 95% confidence interval for θ takes the form

$$\widehat{\theta} \quad \pm \quad t(16, 0.025)\widehat{\sigma}_{\widehat{\theta}}$$

$$4.2605 \quad \pm \quad 2.120(0.9066) \quad \text{or} \quad (2.2385, 6.1825).$$

With 95% confidence we conclude that the total mean height of treated plots that are fertilized is between 2.2385 and 6.1825 higher than the total mean height of treated unfertilized plots.

(f) The difference between the average 5 year heights for plants grown in control sites and those grown in prepared sites can be estimated by

$$\widehat{\theta} = \overline{Y}_{+5+} - \frac{1}{4}(\overline{Y}_{+1+} + \overline{Y}_{+2+} + \overline{Y}_{+3+} + \overline{Y}_{+4+})$$

$$= 9.1136 - \frac{1}{4}(10.9881 + \ldots + 8.7819) = -1.0569.$$

Since the above contrast is a comparison of subplots, the standard error of θ can be found as

$$\widehat{\sigma}_{\widehat{\theta}} = \sqrt{\frac{MS[E]}{2(4)} + \frac{4MS[E]}{4^2(2)(4)}}$$

$$= \sqrt{\frac{5(1.6437)}{32}} = 0.5068,$$

which is the standard error of a linear combination of treatment means. Hence the lower bound becomes

$$\widehat{\theta} - t(16, .10)\widehat{\sigma}_{\widehat{\theta}} = -1.0569 - 1.337(0.5068) = -1.7345.$$

We conclude that with 90% confidence the average 5 year height of plants grown in the control is at most 1.7345 below the average 5 year height of plants grown in prepared sites.

16.11. In order to derive the standard errors in the table in the text, we formalize the assumptions.

$$Y_{ijk} = \mu + \alpha_i + \beta_j + (\alpha\beta)_{ij} + S_{ik} + E_{ijk}, \quad i = 1, \ldots, a; j = 1, \ldots, b; k = 1, \ldots, r,$$

where

μ is a constant,
α_i is a fixed effect, such that $\sum_{i=1}^{a} \alpha_i = 0$,
β_j is a fixed effect, such that $\sum_{j=1}^{b} \beta_j = 0$,
$(\alpha\beta)_{ij}$ is a fixed effect, such that $\sum_{i=1}^{a}(\alpha\beta)_{ij} = 0$ and $\sum_{j=1}^{b}(\alpha\beta)_{ij} = 0$,
S_{ik} are independent $N(0, \sigma_S^2)$,
E_{ijk} are independent $N(0, \sigma^2)$,
S_{ik} and E_{ijk} are mutually independent.

Notationally, we have

$$\overline{Y}_{i++} = \mu + \alpha_i + \overline{S}_{i+} + \overline{E}_{i++}$$
$$\overline{Y}_{+j+} = \mu + \beta_j + \overline{S}_{++} + \overline{E}_{+j+}$$
$$\overline{Y}_{ij+} = \mu + \alpha_i + \beta_j + (\alpha\beta)_{ij} + \overline{S}_{i+} + \overline{E}_{ij+}$$

Using the above conditions and the box in the text we get

$$\text{Var}(\overline{Y}_{i++} - \overline{Y}_{p++}) = \text{Var}(\alpha_i - \alpha_p + \overline{S}_{i+} - \overline{S}_{p+} + \overline{E}_{i++} - \overline{E}_{p++})$$
$$= \text{Var}(\overline{S}_{i+} - \overline{S}_{p+}) + \text{Var}(\overline{E}_{i++} - \overline{E}_{p++})$$
$$= \text{Var}(\overline{S}_{i+}) + \text{Var}(\overline{S}_{p+}) + \text{Var}(\overline{E}_{i++}) + \text{Var}(\overline{E}_{p++})$$
$$= \frac{\sigma_s^2}{r} + \frac{\sigma_s^2}{r} + \frac{\sigma^2}{br} + \frac{\sigma^2}{br}$$
$$= \frac{2[b\sigma_s^2 + \sigma^2]}{br}.$$

190

In Exercise **16.10** it was shown that $MS[M(A)]$ is an unbiased estimator of $b\sigma_s^2 + \sigma^2$, hence plugging in $MS[M(A)]$ for $b\sigma_s^2 + \sigma^2$ gives the desired result.

$$
\begin{aligned}
\text{Var}(\overline{Y}_{+j+} - \overline{Y}_{+m+}) &= \text{Var}(\beta_j - \beta_m + \overline{E}_{+j+} - \overline{E}_{+m+}) \\
&= \text{Var}(\overline{E}_{+j+} - \overline{E}_{+m+}) \\
&= \text{Var}(\overline{E}_{+j+}) + \text{Var}(\overline{E}_{+m+}) \\
&= \frac{\sigma^2}{ar} + \frac{\sigma^2}{ar} \\
&= \frac{2\sigma^2}{ar}.
\end{aligned}
$$

Plugging in $MS[E]$, whose expectation is σ^2, for σ^2 gives the desired result.

$$
\begin{aligned}
\text{Var}(\overline{Y}_{ij+} - \overline{Y}_{im+}) &= \text{Var}(\beta_j - \beta_m + \overline{E}_{ij+} - \overline{E}_{im+}) \\
&= \text{Var}(\overline{E}_{ij+} - \overline{E}_{im+}) \\
&= \text{Var}(\overline{E}_{ij+}) + \text{Var}(\overline{E}_{im+}) \\
&= \frac{2\sigma^2}{r}.
\end{aligned}
$$

Plugging in $MS[E]$, whose expectation is σ^2, for σ^2 gives the desired result.

$$
\begin{aligned}
\text{Var}(\overline{Y}_{ij+} - \overline{Y}_{pj+}) &= \text{Var}(\alpha_i - \alpha_p + \overline{S}_{i+} - \overline{S}_{p+} + \overline{E}_{ij+} - \overline{E}_{pj+}) \\
&= \text{Var}(\overline{S}_{i+} - \overline{S}_{p+}) + \text{Var}(\overline{E}_{ij+} - \overline{E}_{pj+}) \\
&= \text{Var}(\overline{S}_{i+}) + \text{Var}(\overline{S}_{p+}) + \text{Var}(\overline{E}_{ij+}) + \text{Var}(\overline{E}_{pj+}) \\
&= \frac{\sigma_s^2}{r} + \frac{\sigma_s^2}{r} + \frac{\sigma^2}{r} + \frac{\sigma^2}{r} \\
&= \frac{2[\sigma_s^2 + \sigma^2]}{r}.
\end{aligned}
$$

Using the results of Exercise **16.10** it can be shown that $((b-1)MS[E] + MS[M(A)])/b$ is an unbiased estimate of $\sigma_s^2 + \sigma^2$, hence plugging this in gives the desired result. The result for the final comparison in the table takes the identical form as that above, since none of the random components change.

16.13. (a) Using the notation in Box 16.1, we have

$$
Y_{ij} = \mu_{ij} + S_{ij} + E_{ij}, \quad i = 1;, j = 1, 2;
$$

where

$\quad Y_{ij} =$ the cell fluidity measurement for the jth treatment of a particular pair,

$\quad \mu_{ij} =$ the expected cell fluidity measurement for the jth treatment of a particular pair,

$\quad S_{ij} =$ a random quantity that equals the difference between μ_{ij} and the expected cell fluidity measurement for the jth treatment of the particular pair,

$\quad E_{ij} =$ a quantity that equals the amount by which the cell fluidity measurement of the jth treatment for the particular pair differs from the pair's expected cell fluidity measurement.

(b) By the form of the above model, it can be seen that this takes the form of a repeated measures analysis with only one level for the between factor and two levels for the within factor.

(c) Under the assumptions in the text, we have that the variance of S_{ij} is σ_{sj}^2 and that the variance of E_{ij} is σ_j^2. Hence

$$
\begin{aligned}
\text{Var}(Y_{ijk} - Y_{ij'k}) &= \text{Var}(\mu_{ij} - \mu_{ij'} + S_{ijk} - S_{ij'k} + E_{ijk} - E_{ij'k}) \\
&= \text{Var}(S_{ijk} - S_{ij'k} + E_{ijk} - E_{ij'k}) \\
&= \text{Var}(S_{ijk}) + \text{Var}(S_{ij'k}) + 2\text{Cov}(S_{ijk}, S_{ij'k}) \\
&\quad + \text{Var}(E_{ijk}) + \text{Var}(E_{ij'k}) + 2\text{Cov}(E_{ijk}, E_{ij'k}) \\
&= \sigma_{sj}^2 + \sigma_{sj'}^2 + 2\rho_{jj'}\sigma_{sj}^2\sigma_{sj'}^2 + \sigma_j^2 + \sigma_{j'}^2.
\end{aligned}
$$

The sphericity condition states that this variance must be constant for all i, j, j', and k. The above equation does not depend on i nor k and since there only exist two levels of the within factor, we have ($j = 1$, $j' = 2$) or ($j' = 1$, $j = 2$), which imply that the above variance is constant and the model satisfies the sphericity condition.

Under the assumptions in the text we get that

$$
\text{Var}(Y_{ijk}) = \sigma_{sj}^2 + \sigma_j^2.
$$

The homogeneity conditions requires this variance to be constant for all i, j, and k; however, since the variance is a function of j the model does not satisfy the homogeneity condition.

16.15. (a)

$$
\begin{aligned}
Y_{ijkl} &= \mu + \alpha_i + \beta_j + (\alpha\beta)_{ij} + S_{ijl} + \gamma_k + (\alpha\gamma)_{ik} + (\beta\gamma)_{jk} + (\alpha\beta\gamma)_{ijk} + E_{ijkl}, \\
&\quad i = 1, 2, 3;, j = 1, 2; k = 1, 2, 3; l = 1, \ldots, 10;
\end{aligned}
$$

where

$$
\begin{aligned}
Y_{ijkl} &= \text{the measured comprehension of the } l^{th} \text{ child in the } i^{th} \text{ group, of} \\
&\quad \text{Gender } j \text{ on test } k, \\
\mu &= \text{the overall mean comprehension,} \\
\alpha_i &= \text{the fixed effect of the } i^{th} \text{ Group,} \\
\beta_j &= \text{the fixed effect of the } j^{th} \text{ Gender,} \\
(\alpha\beta)_{ij} &= \text{the joint effect of the } i^{th} \text{ Group } j^{th} \text{ Gender,} \\
S_{ijl} &= \text{the random effect of the } l^{th} \text{ Subject in the } i^{th} \text{ Group of Gender } j, \\
\gamma_k &= \text{the fixed effect of the } k^{th} \text{ Test,} \\
(\alpha\gamma)_{ik} &= \text{the joint effect of the } i^{th} \text{ Group and } k^{th} \text{ Test,} \\
(\beta\gamma)_{jk} &= \text{the joint effect of the } j^{th} \text{ Gender and } k^{th} \text{ Test,} \\
(\alpha\beta\gamma)_{ijk} &= \text{the joint effect of the } i^{th} \text{ Group, } j^{th} \text{ Gender and } k^{th} \text{ Test,} \\
E_{ijkl} &= \text{the random error effect.}
\end{aligned}
$$

The usual assumptions about the parameters are also made as in the text.

(b) Since both $\tilde{\epsilon}$ and $\hat{\epsilon}$ are quite close to the true upper limit of ϵ we conclude that the assumption of sphericity is reasonable. Hence, the univariate approaches should be valid for this data.

(c) The completed ANOVA table follows.

Source	df	SS	MS	F_c	p–value
GRP	2	1896.844	948.422	17.06	.0001
Gender	1	432.450	432.450	7.78	.0073
GRP×Gender	2	308.133	154.067	2.77	.0716
Main–plot Error	54	3001.567	55.585	——	——
Test	2	2703.244	1351.622	323.00	.0001
Test×GRP	4	14227.022	3556.756	849.97	.0001
Test×Gender	2	8.133	4.067	0.97	.3817
Test×GRP×Gender	4	14.333	3.583	0.86	.4928
Sub–plot Error	108	451.933	4.185	——	——
Total	179	23043.659			

(d) Based on the above univariate ANOVA table the following conclusions can be made. There is no evidence ($p = 0.4928$) of a three way interaction effect between Test, Group and Gender on comprehension. There is also no evidence ($p = 0.3817, 0.0716$) of a two way interactions between Test and Gender nor Group and Gender on comprehension. There is evidence ($p = 0.0001$) of an interaction effect between Test and Group on comprehension. Finally there is also a significant effect ($p = 0.0073$) on comprehension due to Gender alone.

(e) For the multivariate analysis, the results for the interaction between Group and Gender and the main effect are identical to those above, since these correspond to main effects. The conclusions for the other factors also remain identical although the modified p–values are slightly changed for the Greenhouse-Geisser tests. (The modified p–values for the Huynh-Feldt are identical to those for the univariate analysis.)

(f) The effect being measured by the contrast θ_1 corresponds to a difference between the mean effects of Gender when averaged over both Group and Test.

(g) The contrast based on θ_1 is a comparison of the average of sets of Main–plot effects averaged over the Sub–plot effects. Hence the appropriate standard error of θ_1 corresponds to

$$\sigma_{\widehat{\theta}_1} = \sqrt{\frac{2MS[M(A)]}{3(3)(10)}} = \sqrt{\frac{2(55.585)}{90}} = 1.1114.$$

The estimate for θ_1 can be found by substituting in the observed sample means for their true means. This gives

$$\widehat{\theta}_1 = 3.2111.$$

The confidence interval then becomes

$$\widehat{\theta}_1 \quad \pm \quad t(54, 0.025)\sigma_{\widehat{\theta}_1}$$
$$3.2111 \quad \pm \quad 2.0049(1.1114) \quad \text{or} \quad (0.9829, 5.4393).$$

With 95% confidence we conclude that Gender 2 scores on the average between 0.9829 and 5.4393 higher when averaged over all groups and tests.

(h) The effect being measured by the contrast θ_2 corresponds to a difference between the mean effects of Group A with Group N for Test 1 when averaged over Gender. The effect being measured by the contrast θ_3 corresponds to a a difference between the mean effects of Group B with Group N for Test 1 when averaged over Gender.

(i) Both θ_2 and θ_3 are comparisons of different Sub–plot effects for a fixed Main–plot effect, hence the appropriate standard error corresponds to

$$\widehat{\sigma}_{\widehat{\theta}} = \sqrt{\frac{2\left[(3-1)MS[E] + MS[M(A)]\right]}{6(10)}} = \sqrt{\frac{2(4.185) + 55.585}{30}} = 1.4601.$$

The estimated degrees of freedom for these contrasts is

$$\nu = \frac{\left(1.4601^2\right)^2}{\frac{1}{108}\left[\frac{4.185}{15}\right]^2 + \frac{1}{54}\left[\frac{55.585}{30}\right]^2} = 70.6896 \approx 70.$$

The confidence intervals can then be constructed by estimating θ_2 and θ_3 by plugging in the observed sample means for their true means. This gives

$$\widehat{\theta}_2 = 15.45 \qquad \widehat{\theta}_3 = -3.0.$$

For θ_2 the confidence interval becomes

$$\widehat{\theta}_2 \quad \pm \quad t(70, .025)\widehat{\sigma}_{\widehat{\theta}_2}$$
$$15.45 \quad \pm \quad 1.9944(1.4601) \quad \text{or} \quad (12.538, 18.362).$$

For θ_3 the confidence interval becomes

$$\widehat{\theta}_3 \quad \pm \quad t(70, .025)\widehat{\sigma}_{\widehat{\theta}_3}$$
$$-3 \quad \pm \quad 1.9944(1.4601) \quad \text{or} \quad (-5.912, -.088).$$

With 95% confidence we conclude that the average score for Group A is between 12.538 and 18.362 above the average score for Group N on Test 1 when averaged over Gender. Similarly with 95% confidence we conclude that the average score for Group B is between 0.088 and 5.912 below the average score for Group N on Test 1 when averaged over Gender.